AF173793

Lecture Notes in Physics

Founding Editors
Wolf Beiglböck
Jürgen Ehlers
Klaus Hepp
Hans-Arwed Weidenmüller

Volume 1043

Series Editors
Roberta Citro, Salerno, Italy
Peter Hänggi, Augsburg, Germany
Betti Hartmann , London, UK
Morten Hjorth-Jensen, Oslo, Norway
Maciej Lewenstein, Barcelona, Spain
Satya N. Majumdar, Orsay, France
Luciano Rezzolla, Frankfurt am Main, Germany
Angel Rubio, Hamburg, Germany
Wolfgang Schleich, Ulm, Germany
Stefan Theisen, Potsdam, Germany
James D. Wells, Ann Arbor, MI, USA
Gary P. Zank, Huntsville, AL, USA

The series Lecture Notes in Physics (LNP), founded in 1969, reports new developments in physics research and teaching - quickly and informally, but with a high quality and the explicit aim to summarize and communicate current knowledge in an accessible way. Books published in this series are conceived as bridging material between advanced graduate textbooks and the forefront of research and to serve three purposes:

- to be a compact and modern up-to-date source of reference on a well-defined topic;
- to serve as an accessible introduction to the field to postgraduate students and non-specialist researchers from related areas;
- to be a source of advanced teaching material for specialized seminars, courses and schools.

Both monographs and multi-author volumes will be considered for publication. Edited volumes should however consist of a very limited number of contributions only. Proceedings will not be considered for LNP.

Volumes published in LNP are disseminated both in print and in electronic formats, the electronic archive being available at springerlink.com. The series content is indexed, abstracted and referenced by many abstracting and information services, bibliographic networks, subscription agencies, library networks, and consortia.

Proposals should be sent to a member of the Editorial Board, or directly to the responsible editor at Springer:

Dr Lisa Scalone
lisa.scalone@springernature.com

Balint Radics

Neutrino Physics

A Student's Guide to Simulation

 Springer

Balint Radics 🆔
Universitat Politècnica de Catalunya
Barcelona, Spain

ISSN 0075-8450   ISSN 1616-6361   (electronic)
Lecture Notes in Physics
ISBN 978-3-032-03992-7   ISBN 978-3-032-03993-4   (eBook)
https://doi.org/10.1007/978-3-032-03993-4

This Springer imprint is published by the registered company Springer Nature Switzerland AG
The registered company address is: Gewerbestrasse 11, 6330 Cham, Switzerland

If disposing of this product, please recycle the paper.

To the students who have the patience and curiosity...

Declarations

Competing Interests The author has no competing interests to declare that are relevant to the content of this manuscript.

Contents

Introduction

1

It is nice to know that the computer understands the problem.
But I would like to understand it, too.

Eugene Wigner

Abstract

This chapter outlines the purpose and scope of the book, intended for graduate students in particle physics who wish to take a computational approach to neutrino interactions and flavour oscillations. It also recommends background reading from textbooks to help contextualize the material. Finally, the overall structure of the book is presented, highlighting how each chapter builds towards simulating processes and signatures in neutrino physics.

1.1   Purpose

After almost a hundred years since the neutrino's theoretical inception by Wolfgang Pauli[1] in the 1930s—as a "desparate remedy" to restore energy conservation in β decays of radioactive nuclei—neutrinos have become central players in the field of particle physics. Being the most abundant massive particles in the universe and with sources ranging from the Sun, the Earth's crust, supernovae, cosmic rays, and nuclear reactors to particle accelerators, neutrinos permeate the cosmos, emerging from both natural and man-made experiments. Despite their weak interactions, the intensity of their sources, combined with advances in detector technology and target mass, has enabled routine detection and research in the past few decades.

---

[1] Austrian theoretical physicist W. Pauli (1900–1958).

© The Author(s), under exclusive license to Springer Nature Switzerland AG 2026
B. Radics, *Neutrino Physics*, Lecture Notes in Physics 1043,
https://doi.org/10.1007/978-3-032-03993-4_1

This book serves as a computational physics guide for students interested in this domain. Its aim is to make learning enjoyable and hands-on by working with numbers and code in the context of particle physics, particularly with neutrinos. The motivation stems from years of experience in teaching and working on both large and small experiments, where it is often observed that university curricula lack coding as an integral part of physics education. Additionally, while many excellent textbooks cover the theoretical foundations, they typically provide little guidance on how real experiments operate or how computer simulations of particle scattering or decay can be realized. For an undergraduate or graduate student joining a particle physics experiment, the experience can be both exciting and overwhelming: absorbing vast amounts of information, navigating thousands of lines of legacy code, and understanding detector outputs and data formats. Despite the abundance of information, it can be difficult to distill the essential physical phenomena that lie behind the experiments.

Why focus on neutrinos? There are several reasons. Neutrinos are ultrarelativistic, which simplifies many kinematic calculations. Leading-order results from electroweak perturbation theory often suffice to reproduce experimental outcomes with reasonable qualitative accuracy. Key features of various experiments can be modelled using only two- or three-body phase-space sampling. Furthermore, the phenomenon of three-flavour neutrino oscillation can be described by a set of first-order ordinary differential equations (ODEs) that are readily solvable with modern computational tools on a portable computer. The discovery of neutrino oscillations has elevated the field to the forefront of research, and we are at a historic juncture where dedicated experiments aim to measure several fundamental unknowns with great precision. Next-generation long-baseline oscillation experiments are expected to reach unprecedented sensitivity in the measurement of leptonic CP violation, the determination of the neutrino mass ordering, while also providing stringent tests of the three-flavour oscillation paradigm and conducting searches for new physics. Beyond oscillations, several fundamental open questions remain about neutrinos, such as their absolute mass or their Majorana vs. Dirac nature. Neutrinos also play pivotal roles in many other areas such as Big Bang nucleosynthesis, supernova bursts, radiogenic heating of Earth, dark matter searches, etc.

Another important motivation for this guide is the ubiquity of powerful computing resources. In the 1940s, the first computer, the ENIAC (Electronic Numerical Integrator and Computer), filled a room, weighed almost 50 tons, and could perform only a few thousand operations per second. Today, a portable computer weighs 1.5 kg and offers computational power measured in hundreds of gigaflops.[2] As a result, simulating thousands of neutrino-nucleon interactions or modelling neutrino propagation through the entire Sun can now be done in a few minutes. This book includes examples of both. Any physics student with curiosity and minimal coding experience can reproduce them. Many tools—open-source libraries, high-level programming languages, and numerical solvers—are freely available.

---

[2] 1 gigaflop equals 10^9 floating point operations per second.

Finally, it is important to note that this book is neither a substitute for a standard textbook on particle physics nor a comprehensive guide to the intricacies of modern event generators. It does not aim to cover all aspects or the full complexity of current neutrino physics research. Rather, its goal is to demonstrate how to reproduce and capture key features of some important processes and experiments with a bit of coding—and to have fun doing so.

1.2   Audience and Recommended Resources

This book is intended for graduate students who have taken courses in special relativity and introductory particle physics. Numerous excellent textbooks and advanced resources are available to help readers deepen their understanding. A non-exhaustive list includes:

- D. J. Griffiths, *Introduction to Elementary Particles* (Wiley)
- F. Halzen, A. D. Martin, *Quarks and Leptons* (Wiley)
- D. H. Perkins, *Introduction to High Energy Physics* (Cambridge)
- W. Greiner, B. Müller, *Gauge Theory of Weak Interactions* (Springer)
- A. Rubbia, *Phenomenology of Particle Physics* (Cambridge)
- M. Thomson, *Modern Particle Physics* (Cambridge)
- C. Giunti, C. W. Kim, *Fundamentals of Neutrino Physics and Astrophysics* (Oxford)
- K. Zuber, *Neutrino Physics* (CRC Press)

Additional recommended resources:

- B. Kayser, *Neutrino Physics* in the PDG Review of Particle Physics
- R. Hagedorn, *Relativistic Kinematics* (W. A. Benjamin Inc.)
- F. Suekane, *Neutrino Oscillations* (Springer)
- J. N. Bahcall, *Neutrino Astrophysics* (Cambridge)
- A. C. Phillips, *The Physics of Stars* (Wiley)

The programming language chosen for this book is C/C++. It is general enough to be translated into other languages and can be compiled on most machines. Moreover, it is the native language of the ROOT data analysis framework, which is widely used in the particle physics community. ROOT also integrates naturally with the GNU Scientific Library (GSL), which offers tools for advanced numerical algorithms using real and complex algebra. Some basic programming knowledge is also assumed. The CERN ROOT library offers excellent tutorials and courses for beginners:

- https://root.cern/tutorials/
- https://root.cern/learn/courses/
- https://root.cern/primer/

1.3   Overview of the Book's Structure

The book is structured as follows:

Chapter 2 provides a "Computational Toolbox" that introduces basic units, the ROOT library, and essential numerical methods, such as rejection sampling, numerical integration, and solving ODEs. Minimal example codes are included.

Chapter 3 gives a short reminder of the kinematics in special relativity necessary for understanding the equations used in event generators and interpreting physical constraints in kinematic distributions. It also features examples involving Lorentz boosts, basic phase-space generation, and final-state kinematics for decay and scattering processes.

Chapter 4 introduces the application of rejection sampling to weak decay processes, such as neutron decay, muon decay, and charged pion and kaon decays in accelerator contexts. A section is also dedicated to solar neutrino spectra.

Chapter 5 discusses the fundamentals of neutrino-nucleon scattering, focusing on charged-current quasi-elastic (CCQE) and single-pion resonance (CC RES) processes, using the Llewellyn Smith and Rein-Sehgal models. The chapter illustrates how to calculate cross sections and generate final-state kinematics.

Chapter 6 focuses on neutrino flavour oscillations. It explains how to solve the corresponding complex ODE system and includes examples of neutrino propagation in vacuum and in matter, with constant and varying density. The chapter culminates with a discussion of far detector spectra at the T2K long-baseline experiment, reproducing the electron neutrino appearance spectrum qualitatively. The book concludes with an overview of past, present, and future neutrino experiments.

Computational Toolbox

2

Abstract

This chapter introduces the computational tools used throughout the book to perform numerical calculations. We begin with an overview of natural units commonly used in particle physics and demonstrate how to convert between different unit systems. A brief introduction to the CERN ROOT library illustrates how to generate random numbers, visualize numerical distributions, and randomly sample an arbitrary function using the rejection sampling algorithm. We also show how to use ROOT's built-in numerical integration routines with simple one- and two-dimensional functions and how a simple Monte Carlo integration works. The GSL library is introduced for solving ordinary differential equations (ODEs) in both the real and complex domains. Finally, we provide a short overview of ROOT's built-in physical and mathematical constants, along with references to other well-known publicly available online databases.

2.1 Units

In physics—and likely in many other scientific disciplines—carefully tracking and understanding units is half the battle. When dealing with code running on a computer, there are several reasons why units deserve attention:

- Numerical output from software typically does not include units.
- Quantities encountered in particle physics can be extremely small or large depending on the unit system used.
- Checking results for correct unit usage can help identify errors in computations.

In this book, we will frequently use both natural units (NU) and international system of units (SI, also referred to as the metre-kilogram-second-ampere, or

B. Radics, *Neutrino Physics*, Lecture Notes in Physics 1043,
https://doi.org/10.1007/978-3-032-03993-4_2

MKSA, system) for numerical calculations. Converting between unit systems involves counting powers of fundamental constants. In natural units used in particle physics, we often define a system in which specific constants—namely, the speed of light and the reduced Planck constant—are set to unity: $c = \hbar = 1$. All other quantities can then be expressed in terms of energy.

For example, consider the SI unit of area, which is given as $kg^0 m^2 s^0$. To express this in natural units, we solve a system of equations using the units of energy E $(kg \cdot m^2/s^2)$, $\hbar$ $(kg \cdot m^2/s)$, and c (m/s):

$$kg^0 m^2 s^0 = E^\alpha \hbar^\beta c^\gamma = (kg \cdot m^2 \cdot s^{-2})^\alpha (kg \cdot m^2 \cdot s^{-1})^\beta (m \cdot s^{-1})^\gamma. \qquad (2.1)$$

This yields a system of three equations for the powers α, β, and γ:

$$0 = \alpha + \beta \qquad (2.2)$$
$$2 = 2\alpha + 2\beta + \gamma$$
$$0 = -2\alpha - \beta - \gamma.$$

From the first equation, we have $\alpha = -\beta$. Substituting into the second gives $\gamma = 2$, and from the third, we find $\alpha = -2$. Thus, $\beta = 2$.

If we express energy in GeV (1 GeV = $10^9 \times 1.602 \times 10^{-19}$ J), we obtain

$$1\,GeV^{-2} \cdot (\hbar c)^2 = 1\,GeV^{-2} = 0.389 \times 10^{-31}\,m^2, \qquad (2.3)$$

where the first equality follows from choosing $c = \hbar = 1$.

A similar procedure allows us to derive conversion factors for other quantities. As an exercise, consider the conversion for momentum:

$$1\,GeV^1 \cdot \hbar^0 \cdot c^{-1} = 1\,GeV = 5.343 \times 10^{-19}\,kg \cdot m/s. \qquad (2.4)$$

For practical purposes, a selection of common conversions is summarized in Table 2.1. Additional factors of 10 can be applied to express other units.

Table 2.1 Natural units for physical quantities and conversion factors between MKSA (SI) and natural units for selected examples

Quantity	MKSA units	Natural units	Conversion factor, $x: x \cdot NU \rightarrow SI$
Length	m	$1\,GeV^{-1}$	$\hbar c = 1.973 \times 10^{-16}$
Area	m^2	$1\,GeV^{-2}$	$(\hbar c)^2 = 0.3894 \times 10^{-31}$
Mass	kg	$1\,GeV$	$c^{-2} = 1.782 \times 10^{-27}$
Energy	$kg \cdot m^2/s^2$	$1\,GeV$	1.602×10^{-10}
Momentum	$kg \cdot m/s$	$1\,GeV$	$c^{-1} = 5.343 \times 10^{-19}$
Time	s	$1\,GeV^{-1}$	$\hbar = 6.583 \times 10^{-25}$

In natural units, quantities that cannot be added or subtracted in SI units may be combined without additional conversion factors. For instance, the following two equations describe the same relation, but the simplified form is only valid when using natural units because energy, momentum, and mass share the same unit:

$$E = \sqrt{\mathbf{p}^2 c^2 + m^2 c^4} \rightarrow E = \sqrt{\mathbf{p}^2 + m^2}.$$

As a result, for massless particles, we often write $E = |\mathbf{p}|$ in natural units.

Textbooks sometimes omit constants like c and $\hbar$ in algebraic expressions or retain them for clarity. In this book, I will always indicate the units of numerical values shown in code listings to avoid ambiguity. However, I will not restrict the numerical work to a single unit system. For example, the muon lifetime is typically quoted as $\tau_\mu \simeq 2.2\,\mu$s in SI units (rather than $\tau_\mu \simeq 3.3 \times 10^{18}$ GeV^{-1} in natural units), whereas the muon mass is usually given as $m_\mu \simeq 105$ MeV in natural units (rather than $m_\mu \simeq 1.88 \times 10^{-28}$ kg in SI units). Clearly, whichever unit system allows for simpler numerical representation is preferable.

Some quantities, of course, are dimensionless. In all cases, best practice is to explicitly indicate the assumed units in any code.

2.2 The ROOT Library

ROOT [1] is an object-oriented software library originally developed at CERN. While its primary user base consists of particle physics researchers, it is also used in other fields such as astronomy. ROOT provides a comprehensive ecosystem for data generation, storage, analysis, visualization, and more. This chapter offers a very brief introduction and overview of some of the components of the library that will be used throughout this book.

An excellent starting point for installing the ROOT library can be found at: https://root.cern/install/. ROOT is supported on Linux, macOS, and Windows operating systems. The simplest way to begin is to download a precompiled binary distribution and run the provided installation scripts or executables. Alternatively, ROOT can be installed using the package manager appropriate to your operating system or compiled directly from source.

2.2.1 Interacting with ROOT via the Interactive Shell

Assuming ROOT is already installed on your system, launching the main executable `root` typically opens the interactive shell environment. From this shell, any function or class provided by the ROOT library can be invoked or instantiated.

Code listing 2.1 Starting the ROOT interactive shell from command line

```
$ root
 --------------------------------------------------------------------------
 | Welcome to ROOT 6.28/00                              https://root.cern |
 | (c) 1995-2022, The ROOT Team; conception: R. Brun, F. Rademakers |
 | Built for macosx64 on Feb 03 2023, 14:50:41                           |
 | From tags/v6-28-00@v6-28-00                                           |
 | With Apple clang version 14.0.0 (clang-1400.0.29.202)                 |
 | Try '.help'/'.?', '.demo', '.license', '.credits', '.quit'/'.q' |
 --------------------------------------------------------------------------

root [0]
```

Another advantage of the interactive shell is that it understands C/C++ programming
instructions. This is essential, as most ROOT functionality is accessed through
interactions with C++ classes. In this book, we will primarily use the following
classes:

- `TMath`: A collection of mathematical operations, functions, and constants.
- `TRandom`: A class for generating random numbers.
- `TH1F`, `TH2F`: Classes for creating one- and two-dimensional histograms, useful
  for summarizing statistical distributions (F denotes floating-point precision).
- `TF1`, `TF2`: Classes for defining functions with specified arguments, ranges, and
  other properties.
- `TLorentzVector`: A general-purpose class for storing and manipulating four
  vectors in relativistic kinematics.
- `TGenPhaseSpace`: A utility class for generating n-body phase-space configura-
  tions based on relativistic kinematics.
- `ROOT::Math::Integrator...`: A set of classes for numerically integrating
  functions. These use numerical algorithms from the GNU Scientific Library
  (GSL) [2], accessible via ROOT's `MathMore` module.

The GNU Scientific Library (GSL) is widely available across various operating
systems, with precompiled binary packages readily accessible. Comprehensive
documentation can be found at: https://www.gnu.org/software/gsl/.[1]

To illustrate the use of these ROOT classes, consider the typical sequence of steps
in most scientific analysis:

---

[1] At the time of writing, the author found a useful article with installation instructions for GSL at
the following link: https://solarianprogrammer.com/2020/01/26/getting-started-gsl-gnu-scientific-
library-windows-macos-linux/

1. Random sampling and data collection
2. Calculation of summary statistics
3. Building a model to describe the data

Let us take an example and implement these steps in code. Suppose we are measuring the mass of a particle with a true mass of $m = 105$ MeV (in natural units), but the measurements are smeared due to finite detector resolution. We model this resolution effect as a Gaussian distribution with zero mean and unit variance. The collected data will then consist of continuous real-valued numbers distributed around a mean close to the true particle mass.

We want to visualize this data using a histogram as a summary statistic, and then overlay a continuous model to describe the distribution. ROOT allows us to simulate this entire workflow. We begin by assigning the true particle mass, m, to a variable and then generate random numbers from a Gaussian distribution centred on m. The following lines can be typed directly into the ROOT interactive shell:

Code listing 2.2   Random number generation

```
root [0] float m = 105.0;
root [1] TRandom ran(0);
root [2] for(unsigned int i = 0; i < 5; i++) cout << m + ran.Gaus(0, 1)
↪ << endl;
104.025
105.628
104.35
105.217
103.815
```

Naturally, we want to store the generated values and visualize them using a histogram. To do this, we first create a one-dimensional histogram object of type TH1F with 20 bins and then fill it with the generated values. The resulting histogram can be visualized immediately.

Code listing 2.3   Random number generation and display in a ROOT histogram object

```
root [0] float m = 105.0;
root [1] TRandom ran(0);
root [2] TH1F h("histo", "Histogram of the data", 20, 100, 110);
root [3] for(unsigned int i = 0; i < 1000; i++) h.Fill(m + ran.Gaus(0,
↪ 1));
root [4] h.GetXaxis()->SetTitle("Mass [MeV]");
root [5] h.GetYaxis()->SetTitle("No of events");
root [6] h.Draw("E1");
```

The third line above creates a TH1F histogram object with 20 bins spanning the range from 100 to 110. This means each bin has a width of $10/20 = 0.5$. The histogram then counts the number of entries falling into each bin, with bin ranges defined as $[100+i \cdot 0.5, 100+(i+1) \cdot 0.5]$ for $i = 0, 1, \ldots, 19$. The next line fills the histogram with 1000 random values drawn from a Gaussian distribution centred on the true particle mass. Subsequent lines set the titles of the horizontal and vertical axes and finally display the histogram. The E1 drawing option instructs ROOT to add symmetric error bars to each bin, representing Gaussian uncertainties based on the bin content.

To compare the data with a theoretical model, see code listing 2.5, which repeats the above steps in a standalone format. As before, we generate random numbers and plot them using a histogram. We now also overlay a toy model using the TF1 class. The expression [0] * TMath::Gaus(x, [1], [2], 1) passed to the TF1 object calls ROOT's built-in implementation of the Gaussian probability density function (pdf), evaluated at a point x:

$$f(x; \mu, \sigma) = \frac{1}{\sqrt{2\pi}\sigma} \exp\left(-\frac{(x-\mu)^2}{2\sigma^2}\right), \tag{2.5}$$

where μ (corresponding to [1] in the code) is the expectation value, and σ^2 (with standard deviation [2]) is the variance. The multiplicative factor [0] acts as a normalization constant. The last argument instructs ROOT to normalize the function (include the factor $\frac{1}{\sqrt{2\pi}\sigma}$).

We then plot this model on top of the histogram representing the data. To enhance the clarity of the visualization, we also add a legend to explain the graphical elements. At this stage, the number of commands increases, making it more convenient to write a short script using a text or code editor. This way, we avoid retyping commands interactively line by line. Save the code shown in code listing 2.5 to a file named mycode.C, and execute it from the ROOT command line using:

```
.x mycode.C
```

The output should be similar to the distribution shown in Fig. 2.1.

Code listing 2.4 Executing a code from a file in an interactive ROOT session

```
$ root
root [0] .x mycode.C
```

Code listing 2.5 Code to generate random data and draw together with a model

```
{
    float m = 105.0; // mass [MeV]
```

```cpp
// random number generator with seed 0
TRandom ran(0);

unsigned int nbins = 20; // number of bins
float xmin = 100; // x axis minimum
float xmax = 110; // x axis maximum

// histogram to keep counts of data
TH1F h("histo", "Histogram of the data", nbins, xmin, xmax);

// generate random smearing around the mass
unsigned int nevents = 1000;
for(unsigned int i = 0; i < nevents; i++){
  h.Fill(m + ran.Gaus(0, 1));
}

// set the axis title
h.GetXaxis()->SetTitle("Mass [MeV]");
h.GetYaxis()->SetTitle("No of events");

// draw the data histogram
h.Draw("E1");

// create a model function with parameters
TF1 f("model", "[0]*TMath::Gaus(x, [1], [2], 1)", xmin, xmax);

// set parameters of the model:
// scaling, mean, standard deviation
float scalefactor = (xmax - xmin)/nbins;
f.SetParameters(nevents*scalefactor, m, 1);
f.Draw("same");

// create legend entries for data and model
TLegend leg(0.1,0.7,0.3,0.9);
leg.AddEntry(&h,"Data histogram");
leg.AddEntry(&f,"Model function");
leg.Draw();
}
```

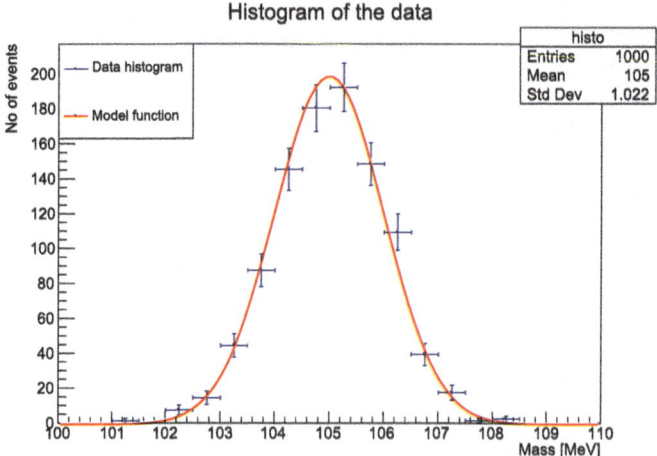

Fig. 2.1 Output of the first ROOT code

2.2.2   Rejection Sampling

The central concept of event generation is the process of *random sampling from a distribution*. This idea is widely used not only in physics but also in statistics. The key insight is that, although the true underlying laws of nature are hidden from us, we can still (i) guess their form and express them using mathematical functions, (ii) design and conduct experiments, and (iii) test theoretical predictions against observations to infer the validity of the theory.

Let us now assume that a theory provides a new mathematical expression predicting how particles behave during interactions. When a particle physics experiment is conducted, the result is a discrete set of decay or scattering events (registered by detectors catching the final-state particles), which represent random samples drawn from nature's true—yet unknown—distribution function. To compare theoretical model predictions with experimental data, we can also generate random samples from the known distributions provided by the model and compare the two.

One simple and popular method to achieve this is called *rejection sampling* [3], also called Von Neumann acceptance-rejection method.[2]

Rejection sampling is a random sampling method that enables generating random numbers that follow the shape of a known probability distribution function, $f(x)dx$. We will use this algorithm frequently in later chapters to simulate particle decays and scattering processes.

As a first example, we demonstrate how to sample randomly from the built-in TMath::Landau function provided by the ROOT library (see Fig. 2.2). This

---

[2] Proposed by Hungarian-American mathematician J. von Neumann (1903–1957).

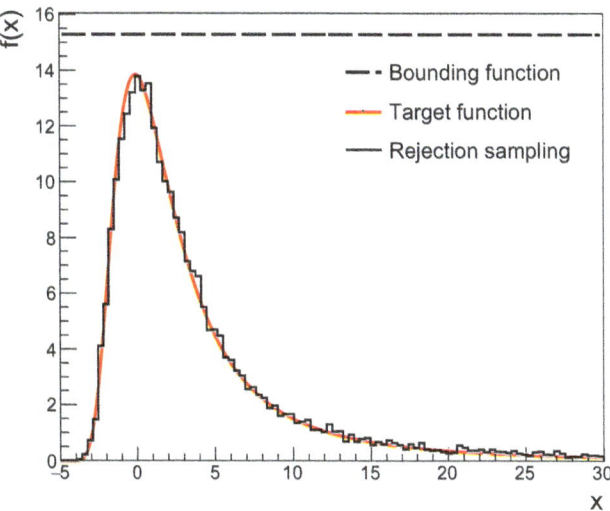

Fig. 2.2 Sampling distribution of x values (black histogram) from the target Landau function (red, solid line) using the rejection sampling method with the bounding function (black, dashed line)

function takes three parameters: an overall multiplicative factor s, the most probable value, mpv, and a scale parameter, σ. Let us use $s = 100$, $mpv = 0.2$, and $\sigma = 1.3$.

To illustrate how the random sampling method works, consider the following steps: Plot the distribution function—in this case, the target being the Landau distribution shown as the red solid line in Fig. 2.2. Then plot a second function $g(x)$—the black dashed line—that bounds the target from above at all points. We then uniformly sample random x values over the domain of interest. For each x, we evaluate $f(x)/(cg(x))$, where c is a scaling factor chosen so that $cg(x) \geq f(x)$ everywhere. Next, we draw a random number α uniformly between 0 and 1 and accept the proposed x value if $\alpha < f(x)/(cg(x))$ at that point. Otherwise, the value is rejected and a new draw is made. Code listing 2.6 below illustrates how to implement this procedure.

Rejection Sampling Algorithm

1. Define the target function $f(x)$ and a bounding function $cg(x)$ such that $cg(x) \geq f(x)$ for all x.
2. Draw a uniformly distributed random number x (a *proposal value*) from the domain of interest, and compute $f(x)/(cg(x))$.
3. Draw a uniformly distributed random number $\alpha \in [0, 1]$.
4. Accept x if $\alpha < f(x)/(cg(x))$; otherwise, reject it and go back to step 2.
5. Repeat to collect enough accepted samples, and create a histogram of the accepted x values.

Let us derive the theoretical value for the *acceptance rate*, $\epsilon = N_{acc}/N_{gen}$, the ratio of the number of samples accepted and that generated. In our algorithm, we uniformly sample $x \in [a, b]$ and evaluate $f(x) < c \cdot g(x)$ for all x. Then, if we choose a uniform proposal function $c \cdot g(x) = c \cdot f_{max}$, we effectively draw a uniform $\alpha \in [0, c \cdot f_{max}]$ and accept if $\alpha < f(x)$. The probability of accepting is the fraction of the rectangle $[a, b] \times [0, c \cdot f_{max}]$ that lies under the curve $f(x)$:

$$\epsilon = \frac{\int_a^b f(x)dx}{\int_a^b c \cdot g(x)dx} = \frac{\int_a^b f(x)dx}{c \cdot f_{max} \cdot (b-a)}. \tag{2.6}$$

The integral $\int_a^b f(x)dx$ is the normalization of the target distribution over the volume, which can be approximated as

$$\int_a^b f(x)dx = \epsilon \cdot c \cdot f_{max} \cdot (b-a) \approx \left(N_{acc}/N_{gen}\right) \cdot c \cdot f_{max} \cdot (b-a). \tag{2.7}$$

Hence, we should be able to automatically normalize our accepted events to our the target by scaling them with a factor $\frac{N_{acc}}{N_{gen}} \cdot c \cdot f_{max} \cdot (b-a)$, which is what we do in the example 2.6.

Code listing 2.6 Rejection sampling example

```
{
    // Initialize random number generator and seed
    TRandom * ran = new TRandom(12343);
    // Histogram for drawing the distribution
    TH1F * hSample = new TH1F("hSample", "", 100, -5, 30);

    // Limits
    double xmin = -5;
    double xmax = 30;

    // The target function
    TF1 * fTarget = new
↪    TF1("fTarget","[0]*TMath::Landau(x,[1],[2],1)",xmin,xmax);
    fTarget->SetParameters(100.0, 0.2,1.3);
    fTarget->SetNpx(1e+03);


    // The envelope function that bounds the target
    // is just a constant function
    double fmax = fTarget->GetMaximum();
    double c = 1.1;

    // The bounding function
```

```cpp
TF1 * fEnvelope = new TF1("fEnvelope","[0]",xmin,xmax);
fEnvelope->SetParameters(c*fmax);

unsigned int Nev = 1e+05;
unsigned int Nacc = 0;
double x = 0;
double fx = 0;
// Draw random numbers in a loop
for(unsigned int iEv = 0; iEv < Nev; iEv++){
  // Generate a random number between -5 and 30
  x = ran->Rndm()*35-5;

  // Evaluate our target function
  fx = fTarget->Eval(x);

  // Throw a random number to decide to accept it or not
  bool accept = (ran->Rndm() < fx/(c*fmax) );

  if(accept){
    Nacc +=1 ;
    hSample->Fill(x);
  }
}

// Calculate the acceptance rate
double epsilon = (double)Nacc/Nev;
// Get the true target integral
double integral_target = fTarget->Integral(xmin, xmax);
// Get the estimated target integral from accepted events
double integral_target_est = epsilon*c*fmax*(xmax-xmin);

std::cout << "Acceptance rate: " << epsilon << std::endl;
std::cout << "Theoretical acceptance rate: " <<
↪   integral_target/(c*fmax*(xmax-xmin)) << std::endl;
std::cout << "Target integral: " << integral_target << std::endl;
std::cout << "Target est. integral: " << integral_target_est <<
↪   std::endl;

// Normalize the accepted samples to the target
hSample->Scale(epsilon*c*fmax*(xmax-xmin)/hSample->Integral("width"));

// Draw the target and envelope functions, and set style
fEnvelope->Draw();
fEnvelope->GetYaxis()->SetRangeUser(0, c*fmax*1.05);
```

```
fEnvelope->SetLineStyle(9);
fEnvelope->SetLineColor(kBlack);
fEnvelope->GetXaxis()->SetTitle("x");
fEnvelope->GetYaxis()->SetTitle("f(x)");
fTarget->Draw("same");

// Draw the accepted random samples
hSample->Draw("histsame");
}
```

Code listing 2.7 Executing the code 2.6 in an interactive ROOT session

```
$ root
root [0] .x RejSamp.C
Acceptance rate: 0.17651
Theoretical acceptance rate: 0.177769
Target integral: 95.1096
Target est. integral: 94.4361
```

Saving the code in listing 2.6 in a file named RejSamp.C and executing it yield random samples whose distribution closely follows the target Landau distribution see Fig. 2.2. During the execution of the algorithm, points in the target domain are accepted with a probability proportional to their value under the distribution. Points corresponding to values of the distribution where $f(x)/(cg(x)) \approx 1$ are more likely to be accepted (they have a higher weight), while those where $f(x)/(cg(x)) \ll 1$ are more likely to be rejected (lower weight). In this case, however, no real weights are necessary; the sampling automatically weights the points according to the distribution. Later we will see other examples where weights are needed.

This method comes with a trade-off. The code reports an acceptance rate of around 17.7%—due to our use of a simple constant bounding function. This choice means that many proposed points, especially those in regions far from the peak, will be rejected. Such inefficiency becomes more pronounced when the target distribution has narrow peaks or long tails.

Despite this inefficiency, rejection sampling remains a robust and widely applicable method, particularly when the target distribution is complicated. The core idea is to find the maximum value of the function—either in one dimension or in slices of higher-dimensional spaces—and to use it as a bound for comparison. However, in higher dimensions, $n_{\mathrm{dim}} > 5\text{–}10$, this approach suffers from the *curse of dimensionality*, making it less practical as dimensionality increases.

Another aspect of rejection sampling is trying to estimate the definite integral of a function using sampling in a domain via an event generator (which we will meet later), and generators may also provide weights $w(x)$ for each x. In the case of rejection sampling, we have derived already the integral formula in Eq. 2.7. In general, for the choice of a uniform proposal function $c \cdot f_{max}$ and *assuming uniform sampling of* x, we can approximate the integral of $f(x)$ using

$$\int_a^b f(x)dx \approx \left(N_{acc}/N_{gen}\right) \cdot (c \cdot f_{max}) \cdot \text{vol}, \qquad (2.8)$$

where N_{acc} and N_{gen} are the accepted and generated events, respectively, in the rejection sampling algorithm, $c \cdot f_{max}$ is a uniform proposal function bounding the target function $f(x)$ for all x, and vol is the volume of the finite (possibly multidimensional) domain.

In case the bounding function $c \cdot f_{max}(x)$ is different for each x, the formula is slightly modified to the following:

$$\int_a^b f(x)dx \approx \frac{\text{vol}}{N_{gen}} \sum_i^{N_{acc}} (c \cdot f_{max}(x_i)). \qquad (2.9)$$

In the case of events generated with phase-space weights combined together with rejection sampling, the picture is more complicated. It is often not guaranteed either that x (or the relevant variable) has a flat distribution in the domain as provided by the event generator. If it is not flat, then the sampling is not uniform in the domain, and the integral is more complicated to evaluate (in addition to the event weights). A possible approach is to generate *unweighted events* by applying rejection sampling using the event weights. We will meet such cases later. However, in this case, the integral of a function $f(x)$ can be estimated as

$$\int_a^b f(x)dx \approx \frac{1}{N_{acc}} \sum_i^{N_{acc}} \frac{f(x_i)}{\rho(x_i)}, \qquad (2.10)$$

where N_{acc} is the accepted events by the rejection sampling algorithm, and $f(x_i)/\rho(x_i)$ allows one to correct for the nonuniform sampling of x. Here $\rho(x)$ is the (nonuniform) marginal distribution of the phase-space density over the variable x, the PDF of which may be estimated by filling a histogram with the values of the generated x and normalizing it to unity. Since $f(x)$ is sampled nonuniformly over x, the procedure above divides out this effect and allows estimation of the integral as a simple Monte Carlo average.

2.2.3 Numerical Integration in 1D

Even when the functions we aim to sample are complex, ROOT provides powerful tools for numerical integration, leveraging built-in routines from the GNU Scientific Library [2]. As an illustrative example, consider evaluating the well-known integral that yields the Euler-Mascheroni constant,[3] whose value is $\gamma = 0.577215664901532\ldots$. This constant appears in many areas of physics and mathematics. Its value was originally discovered as the difference between the harmonic series and the natural logarithm:

$$\gamma_n = \left(\sum_{k=1}^{n} \frac{1}{k}\right) - \ln(n) = H_n - \ln(n), \tag{2.11}$$

which converges to γ as $n \to \infty$ (even though both H_n and $\ln(n)$ are divergent). The following definite integral gives its negative value directly:

$$\int_0^{\infty} e^{-x} \ln(x)\, dx = -\gamma. \tag{2.12}$$

To evaluate this integral numerically, we can write a simple program (see Code listing 2.8). We first define a standalone function `myfunction(const double x)` that contains the mathematical expression under the integral sign. Since the integrand is one dimensional, this function takes a single argument, x.

In the main function `mycode()`, we begin by creating a special ROOT object `ROOT::Math::Functor1D`, which allows us to pass the external function to the integrator. We then instantiate a `ROOT::Math::Integrator` object, specifying the numerical algorithm to use. ROOT supports various integration algorithms such as `kGAUSS`, `kADAPTIVE`, `kADAPTIVESINGULAR`, `kNONADAPTIVE`, and others.

Next, we define the lower and upper limits of integration. From Eq. 2.12, it is clear that the lower limit is zero. However, setting the upper limit to infinity is not feasible in numerical code.[4] Instead, we choose a sufficiently large upper limit such that the exponential term e^{-x} becomes negligible beyond that point. This is justified because e^{-x} decays rapidly, while $\ln(x)$ increases very slowly. For example, $e^{-100} \simeq 3 \times 10^{-44}$, while $\ln(100) \simeq 4.6$, making the contribution to the integral beyond $x = 100$ effectively zero.

Note that in ROOT, `TMath::Log(x)` refers to the natural logarithm, while `TMath::Log10(x)` denotes the base-10 logarithm. Finally, we evaluate the integral by calling `Integral(range_min, range_max)`, store the result in a variable, and print it to the console with a precision of 10 significant digits. The code in listing 2.8 can be executed in a ROOT session as before, by saving the source in a file named `mycode.C` and running it with the familiar command:

```
.x mycode.C
```

---

[3] Named after Swiss mathematician L. Euler (1707–1783) who discovered it, and Italian mathematician L. Mascheroni (1750–1800). It is a curious fact that, at the time of writing this manuscript, it remains unknown whether γ is an irrational number or whether it is transcendental.

[4] In symbolic languages such as Mathematica or SymPy, it is possible to specify an infinite upper limit.

Code listing 2.8 One-dimensional numerical integral example

```
double myfunction(const double x){
  // the integrand
  double value = TMath::Exp(-x)*TMath::Log(x);

  return value;
}
void mycode(){
  // ROOT object needs the address of 'myfunction'
  ROOT::Math::Functor1D func1D(&myfunction);
  ROOT::Math::Integrator
↪   integrator(ROOT::Math::IntegrationOneDim::kADAPTIVE);
  integrator.SetFunction(func1D);

  double range_min = 0;
  double range_max = 100;

  double val = integrator.Integral(range_min,range_max);

  std::cout << std::setprecision(10) << val << std::endl;

}
```

Code listing 2.9 Executing the code in listing 2.8 from a file in an interactive ROOT session

```
$ root
root [0] .x mycode.C
-0.5772156649
```

2.2.4   Numerical Integration in 2D

Numerically integrating multidimensional functions in ROOT is conceptually similar to the one-dimensional case and requires only minor changes in the code. The primary difference is that the integrand must now accept multiple arguments corresponding to the function's input dimensions.

As an example, we will estimate the integral of the one-dimensional Gaussian function, e^{-x^2}, by rewriting its square as a two-dimensional integral. This two-dimensional form can be evaluated analytically for comparison with the numerical result.

$$\left(\int_{-\infty}^{\infty} e^{-x^2}dx\right)^2 = \int_{-\infty}^{\infty} e^{-x^2}dx \int_{-\infty}^{\infty} e^{-y^2}dy \qquad (2.13)$$

$$= \int_{-\infty}^{\infty}\int_{-\infty}^{\infty} e^{-(x^2+y^2)}dxdy.$$

Now, let us switch to polar coordinates, where $x^2 + y^2 = r^2$ and the Jacobian of the transformation gives $dx\,dy = r\,dr\,d\theta$. Adjusting the limits of integration accordingly, r ranges from 0 to ∞, and θ from 0 to 2π. It then follows that

$$\int_{-\infty}^{\infty}\int_{-\infty}^{\infty} e^{-(x^2+y^2)}dxdy = \int_{0}^{2\pi}\int_{0}^{\infty} e^{-r^2}rdrd\theta. \qquad (2.14)$$

We will instruct ROOT to numerically evaluate the right-hand side of Eq. 2.14. While it is also possible to directly integrate the left-hand side of Eq. 2.13 numerically, using the polar form allows us to verify the analytical derivation. Therefore, we proceed with the polar coordinate version.

$$\int_{0}^{2\pi}\int_{0}^{\infty} e^{-r^2}rdrd\theta = 2\pi\int_{0}^{\infty} e^{-r^2}rdr = 2\pi\int_{-\infty}^{0}\frac{1}{2}e^{s}ds = \pi\int_{-\infty}^{0} e^{s}ds$$

$$= \pi\left[e^0 - e^{-\infty}\right] = \pi. \qquad (2.15)$$

In the third step, we introduced a change of variables: $s = -r^2$, which implies $ds = -2r\,dr$. The limits of the definite integral must also be adjusted accordingly. Code listing 2.10 shows the implementation of the numerical evaluation.

Code listing 2.10 Two-dimensional numerical integral example

```
double myfunction(const double * x){

    // map the values of the array 'x' to the variables
    double theta = x[0];
    double r = x[1];
    // the integrand
    double value = TMath::Exp(-r*r)*r;

    return value;
}

void mycode(){

    // options for the numerical integral
    ROOT::Math::IntegratorMultiDimOptions opt;
    opt.SetNCalls(1e+07);
```

```
// ROOT objects needs the address of 'myfunction'
ROOT::Math::Functor func2D(&myfunction, 2);
ROOT::Math::IntegratorMultiDim
↪   integrator(ROOT::Math::IntegrationMultiDim::kVEGAS);
integrator.SetFunction(func2D);
integrator.SetOptions(opt);

std::cout << "Number of calls: " << integrator.Options().NCalls() <<
↪   std::endl;

double r_min = 0;
double r_max = 100;
double theta_min = 0.0;
double theta_max = 2*TMath::Pi();

double range_min[2] = {theta_min, r_min};
double range_max[2] = {theta_max, r_max};

double val = integrator.Integral(range_min,range_max);

std::cout << std::setprecision(10) << val << std::endl;

}
```

Code listing 2.11 Executing the code in listing 2.10 from a file in an interactive ROOT session

```
$ root
root [0] .x mycode.C
3.14159282
```

The result will differ slightly from the exact value of $\pi = 3.141592654\ldots$. In this example, we used the kVEGAS numerical integration method, a popular Monte Carlo algorithm. It estimates the integral by randomly sampling the integrand many times and computing the average. The precision of the result improves as the number of samples increases. To control this, we explicitly set the number of function calls in the program using opt.SetNCalls(1e+07). By default, ROOT uses a smaller number of calls, which results in lower precision. The reader is encouraged to experiment by increasing the number of calls to observe how closely the computed result approaches the true value of π.

ROOT also supports other integration methods, including kADAPTIVE, kMISER, and kPLAIN. For each method, it is important to adjust the integration options appropriately to ensure accurate results.

2.2.5 Monte Carlo Integration

Following the spirit of the previous subsection, we now implement the Monte Carlo integration technique using a simple example. This method estimates the integral of a function $f(x)$ using the following approximation:

$$I = \int_a^b f(x)\,dx \quad \longrightarrow \quad \langle I \rangle \approx \langle f \rangle \int_a^b dx \simeq \frac{V}{N} \sum_{i=1}^N f(x_i), \qquad (2.16)$$

where $V = \int_a^b dx = b - a$ is the volume (or length, in one dimension) of the integration domain, and $\langle f \rangle$ is the average value of the function $f(x)$, estimated by sampling:

$$\langle f \rangle \simeq \frac{1}{N} \sum_{i=1}^N f(x_i). \qquad (2.17)$$

Since the estimate is based on a finite number N of random samples, it comes with a statistical uncertainty. An empirical estimate of the variance, $\hat{s}$, in $\langle f \rangle$ can be computed as

$$\hat{s} = \frac{1}{N-1} \left(\sum_{i=1}^N f(x_i)^2 - N\langle f \rangle^2 \right). \qquad (2.18)$$

This in turn gives an empirical estimate of the variance in the integral estimate $\langle I \rangle$:

$$\mathrm{Var}(\langle I \rangle) = \frac{V^2}{N} \hat{s}. \qquad (2.19)$$

As an example, let us estimate the definite integral of the *humps function* (shown in Fig. 2.3) over the interval $[0, 2]$, using Monte Carlo integration with $N = 10^7$ samples. We will also report the estimated statistical error of the result. An example implementation is provided in code listing 2.12.

The function to be integrated is given by

$$f(x) = \frac{1}{(x - 0.3)^2 + 0.01} + \frac{1}{(x - 0.9)^2 + 0.04} - 6. \qquad (2.20)$$

The exact analytical value of the integral is $I = 29.3262$. This can be verified using the symbolic language tool WolframAlpha with the query:

```
integrate 1/((x-0.3)^2 + 0.01) + 1/((x-0.9)^2 + 0.04) - 6 dx
from 0 to 2
```

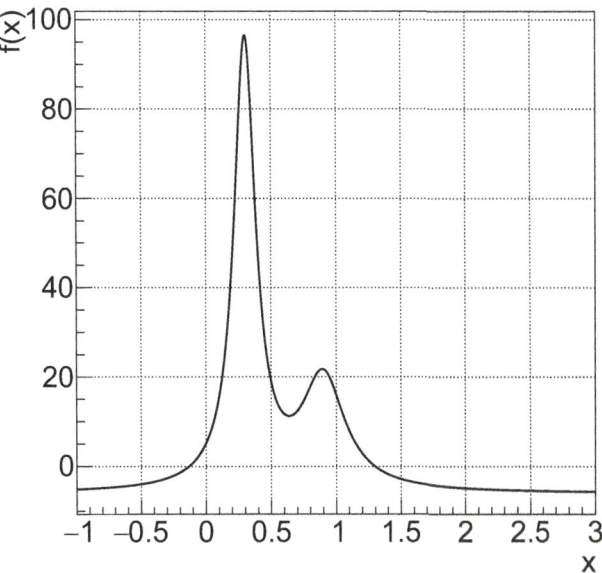

Fig. 2.3 The humps function

Code listing 2.12   Monte Carlo integration of the humps function

```cpp
#include <TRandom2.h>
#include <TMath.h>
#include <TF1.h>
#include <TCanvas.h>
#include <iostream>
#include <iomanip>

// The humps function
double humps(double x){
  return 1.0/(TMath::Power(x-0.3,2) + 0.01) + 1.0/(TMath::Power(x-0.9,2)
↪    + 0.04) - 6;
}

void mycode(){
  // Random number generator
  TRandom2 * ran = new TRandom2();

  // Variables
  double I = 0;
  double s = 0;
```

```
int N = 10000000;
double xmin = 0.0;
double xmax = 2.0;
double x = 0;
double fx = 0;
double vol = xmax - xmin;
double sum_fx = 0;
double sum_fx2 = 0;

// Random sampling
for(int i = 0; i < N; i++){
  x = ran->Uniform(xmin, xmax);
  fx = humps(x);
  sum_fx += fx;
  sum_fx2 += fx*fx;
}

// The estimate mean value of the integral
I = vol*sum_fx/(double)N;

// Estimate the variance
double mean_f = sum_fx / (double)N;
double s_f2 = ((sum_fx2 / N) - mean_f*mean_f) * ((double)N/(N-1));
double var = (vol*vol/N) * s_f2;

  std::cout << "Result:" << std::setprecision(5) << I << " +- " <<
↳  TMath::Sqrt(var) << std::endl;

// Draw the function
TCanvas * c = new TCanvas("c", "c", 1);
c->SetGridx();
c->SetGridy();
TF1 * f = new TF1("humps", "1.0/((x-0.3)*(x-0.3)+0.01) +
↳  1.0/((x-0.9)*(x-0.9)+0.04) - 6", -1, 3);
f->SetNpx(1000);
f->Draw();
}
```

Code listing 2.13 Executing the code in listing 2.12 from a file in an interactive ROOT session

```
$ root
root [0] .x mycode.C
Result:29.326 +- 0.014879
```

The result is reported to five significant digits. Our estimated value of the integral is close to the exact result, with a statistical uncertainty of $\sim 0.05\%$.

A robust—albeit computationally intensive—method for cross-checking the estimated Monte Carlo uncertainty is known as *empirical error estimation*. This approach involves repeating the Monte Carlo integration multiple times (using different random number seeds) and computing the averages and variance of the averages from the resulting values. It is the same Monte Carlo estimation method as the one used before, but repeated on independent samples.

An alternative approach is the so-called *bootstrap resampling* method [4, 5]. Bootstrap is a widely used technique to estimate properties of the sampling distribution of a parameter or other statistic by repeated resampling *with replacement* from an empirical distribution. The original empirical distribution (the observed *data*) is typically drawn from a (unknown) population distribution. Often drawing millions of samples from a population is impossible; therefore, the idea behind the bootstrap is to draw repeated samples from an estimate of the population (the empirical distribution, a single sample). In case the population is very large (even infinite) compared to the sample size, estimations under sampling without replacement are approximated by estimations under sampling with replacement.[5] Since the bootstrap distribution is sampled from the empirical distribution, the bootstrap distribution will be centred at the observed sample statistic, not the population parameter, e.g. at $\bar{x}$ (where $\bar{x}$ could be the mean of an observed dataset, a Monte Carlo estimation of an integral, etc.), and not μ (where μ is the true, unknown parameter or integral value). It means that we do not use the bootstrap to get better estimates of μ. For example, we cannot use the bootstrap to improve on $\bar{x}$; no matter how many bootstrap samples we take, they are always centred at $\bar{x}$, not μ. Instead, the bootstrap is used to tell how accurate the original estimate is by repeatedly sampling the empirical distribution several times.

Therefore, once we generated an initial finite Monte Carlo sample, $f(x_i)$, the values are kept and resampled *with replacement* many times to estimate the sample variance. The method reuses the values from a *single* sample to estimate the uncertainty of that sample without the need for generating new data. The results from these two approaches are illustrated in Fig. 2.4. The empirical method (shown as a solid histogram) draws 10^7 independent samples one hundred times, each time recalculating the integral, and using the one hundred integral values to capture the natural spread due to statistical fluctuations. The bootstrap method (shown as a dashed histogram) resamples a *single sample* of 10^7 function values with replacement one hundred times and estimates the sample uncertainty from the distribution of the resampled means. The Monte Carlo estimate of the integral from the original sample is indicated by a vertical red line in the figure. Numerical results

---

[5] Suppose that, in a population of 10000 people, 3000 have blue eyes and 7000 have eyes of other colour. If we randomly select 10 individuals with replacement, the probability that 3 of them have blue eyes is $\binom{10}{3}(3000/10000)^3(7000/10000)^7 \approx 0.26683$. If the selection is made without replacement, the probability is $\binom{3000}{3}\binom{7000}{7}/\binom{10000}{10} \approx 0.26696$.

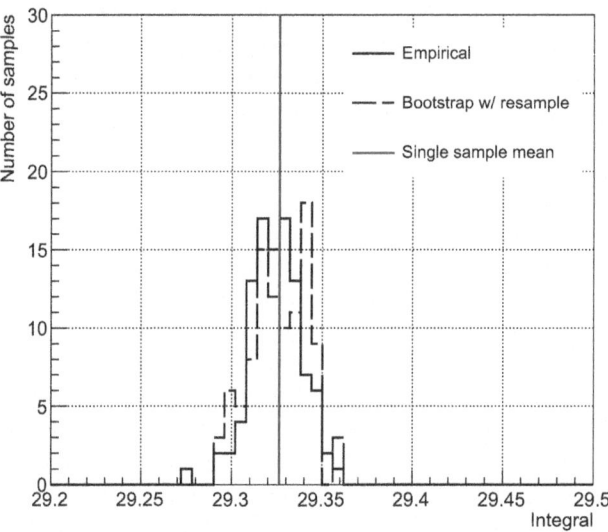

Fig. 2.4 Comparison of empirical (solid histogram) and bootstrap resampling (dashed histogram) estimation of the uncertainty of the Monte Carlo integral, showing also the estimated mean (vertical, red line) from a single Monte Carlo sample

Table 2.2 Comparison of Monte Carlo integral and uncertainty with empirical and bootstrap resampling estimations

Quantity	Mean	Uncertainty
Single sample mean	29.326	0.014879
Empirical	29.325	0.014363
Bootstrap resampling	29.326	0.015990

for a representative example are presented in Table 2.2. The values of the estimated uncertainty agree well in all three cases.

2.3   Solving Ordinary Differential Equations with GSL

This section provides a brief introduction to numerically solving differential equations as initial-value problems. The technical aspects are summarized here so that, in Chap. 6, the physical discussion can proceed without being encumbered by implementation details.

2.3.1   ODE Problem in the Real Domain: Harmonic Oscillator

In addition to random sampling and numerical integration of functions, another important computational tool is the ability to solve initial-value problems for differential equations. In this section, we focus on using the GSL to integrate single ordinary differential equations (ODEs) or systems of ODEs.

A widely used and robust approach is the Runge-Kutta method,[6] which belongs to the class of higher-order Taylor methods for numerically solving ODEs. Consider the general problem:

$$\frac{d\mathbf{y}}{dt} = f(t, \mathbf{y}(t)), \quad \mathbf{y}(t_0) = \mathbf{y}_0, \tag{2.21}$$

where $\mathbf{y}(t) = (y_1(t), y_2(t), \ldots, y_n(t))^T$ is a vector of dependent variables, $t \in [a, b]$ is the independent variable (often interpreted as time), and $f : [a, b] \times \mathbb{R}^n \to \mathbb{R}^n$.

To approximate the solution from a known value $\mathbf{y}(t_n)$ to the next value $\mathbf{y}(t_{n+1})$ with step size $h = (b - a)/M$, where M is the number of discrete time steps such that $t_j = a + jh$, the fourth-order Runge-Kutta method uses the following scheme:

$$\mathbf{y}(t_{n+1}) = \mathbf{y}(t_n) + \frac{1}{6}(k_1 + 2k_2 + 2k_3 + k_4), \tag{2.22}$$

$$k_1 = hf(t_n, \mathbf{y}(t_n)),$$

$$k_2 = hf\left(t_n + \frac{h}{2}, \mathbf{y}(t_n) + \frac{1}{2}k_1\right),$$

$$k_3 = hf\left(t_n + \frac{h}{2}, \mathbf{y}(t_n) + \frac{1}{2}k_2\right),$$

$$k_4 = hf(t_n + h, \mathbf{y}(t_n) + k_3).$$

There are also adaptive versions of this method, such as the Runge-Kutta-Fehlberg algorithm, which includes automatic step-size control. The GSL library provides several such algorithms and associated interfaces. For general n-dimensional systems, GSL also requires the Jacobian matrix—i.e. the partial derivatives of each function component with respect to each dependent variable:

$$J_{ij} \equiv \frac{\partial f_i(t, \mathbf{y}(t))}{\partial y_j}. \tag{2.23}$$

As a concrete example, consider the classical harmonic oscillator equation, $\ddot{x} + \omega^2 x = 0$. We can reduce this second-order equation to a system of first-order equations by defining $y_1 = x$ and $y_2 = \dot{x}$. The system then becomes

$$\dot{y}_1 = y_2 \equiv f_1,$$

---

[6] Developed by German mathematicians C. Runge (1856–1927) and M. W. Kutta (1867–1944).

$$\dot{y}_2 = -\omega^2 y_1 \equiv f_2.$$

This can be written in vector form using a column-vector notation, which is also convenient for implementation in code:

$$\frac{d}{dt}\begin{pmatrix} y_1 \\ y_2 \end{pmatrix} = \begin{pmatrix} y_2 \\ -\omega^2 y_1 \end{pmatrix}. \tag{2.24}$$

The corresponding Jacobian matrix is

$$J_{ij} = \begin{pmatrix} \frac{\partial f_1}{\partial y_1} & \frac{\partial f_1}{\partial y_2} \\ \frac{\partial f_2}{\partial y_1} & \frac{\partial f_2}{\partial y_2} \end{pmatrix} = \begin{pmatrix} 0 & 1 \\ -\omega^2 & 0 \end{pmatrix}. \tag{2.25}$$

At this point, the only remaining task is to specify the initial conditions and the parameter ω^2. For this example, we choose $y_1(t_0) = 1.0$, $y_2(t_0) = 0.0$, and $\omega^2 = 1$. With these definitions, we have all the necessary ingredients to implement the numerical solution in code. See Code listing 2.14 for an example implementation.

Code listing 2.14 Numerical code for the harmonic oscillator ODE using a fixed step-size fourth-order Runge-Kutta algorithm with GSL

```
#include <TFile.h>
#include <TGraph.h>
#include <TCanvas.h>
#include <TStyle.h>
#include <TAxis.h>

#include <stdio.h>
#include <gsl/gsl_errno.h>
#include <gsl/gsl_matrix.h>
#include <gsl/gsl_odeiv2.h>

// Define the right-hand side of the system of ODE
int func (double t, const double y[], double f[], void *params){
  double w_sq = *(double *)params;
  f[0] = y[1];// dx/dt = v = y[1]
  f[1] = -w_sq*y[0];// dv/dt = d2x/dt^2 = -w_sq*x = -w_sq*y[0];
  return GSL_SUCCESS;
}

// The Jacobian matrix
int jac (double t, const double y[], double *dfdy, double dfdt[], void
↪  *params){
  // Pass the omega^2 parameter
  double w_sq = *(double *)params;
```

```cpp
  // Represent the Jacobian as a matrix
  gsl_matrix_view dfdy_mat
    = gsl_matrix_view_array (dfdy, 2, 2);
  gsl_matrix * m = &dfdy_mat.matrix;
  gsl_matrix_set (m, 0, 0, 0.0);
  gsl_matrix_set (m, 0, 1, 1.0);
  gsl_matrix_set (m, 1, 0, -w_sq);
  gsl_matrix_set (m, 1, 1, 0.0);
  // Set the time derivatives of the right-hand side function f
  dfdt[0] = 0.0;
  dfdt[1] = 0.0;
  return GSL_SUCCESS;
}


// The main code execution
int main (void){

  // The omega^2 parameter
  double w_sq = 1.0;
  // Define the ODE system
  gsl_odeiv2_system sys = { func, jac, 2, &w_sq };

  // Create a GSL ODE driver object with step size 1e-03,
  // and absolute and relative error of 1e-08.
  // Note that the Fourth Order Runge-Kutta algorithm is defined here
  gsl_odeiv2_driver *d =
    gsl_odeiv2_driver_alloc_y_new (&sys, gsl_odeiv2_step_rk4,
                                   1e-3, 1e-8, 1e-8);

  // Set initial values
  double t = 0.0; // time zero
  double y[2] = { 1.0, 0.0 }; // initial x and v
  int i, s;

  // Save the intermediate step values
  std::vector<double> y0_v, y1_v, t_v;
  unsigned int nsteps = 0;

  // Take nsteps steps in a loop
  for (i = 0; i < 1000; i++)
    {
      // Evolve the ODE with step size 1e-03 for 10 steps
      s = gsl_odeiv2_driver_apply_fixed_step (d, &t, 1e-3, 10, y);
```

```cpp
      if (s != GSL_SUCCESS)
        {
          printf ("error: driver returned %d\n", s);
          break;
        }

      // Print the intermediate step values to screen
      printf ("%.5e %.5e %.5e\n", t, y[0], y[1]);

      // Save the intermediate step values
      t_v.push_back(t);
      y0_v.push_back(y[0]);
      y1_v.push_back(y[1]);
      nsteps++;

    }


  gsl_odeiv2_driver_free (d);

  // Create a ROOT output file to draw the result in graphs
  // onto a canvas object
  TFile * ofile = new TFile("out.root", "RECREATE");
  // The canvas
  TCanvas *c1 = new TCanvas("c1", "c1",10,65,700,500);
  gStyle->SetOptTitle(0); // no titles
  c1->Divide(2,1); // divide the canvas to pads
  c1->cd(1); // draw on pad 1


  // The graphs holding the result plots
  // x versus time
  TGraph * gr_x_t = new TGraph(nsteps, &t_v[0], &y0_v[0]);
  gr_x_t->SetName("Graph_x_vs_t");
  gr_x_t->GetXaxis()->SetTitle("t");
  gr_x_t->GetYaxis()->SetTitle("x");
  gr_x_t->SetTitle("Graph of x and v vs t");
  gr_x_t->Draw("AP");

  // v versus time
  TGraph * gr_v_t = new TGraph(nsteps, &t_v[0], &y1_v[0]);
  gr_v_t->SetName("Graph_v_vs_t");
  gr_v_t->GetXaxis()->SetTitle("t");
  gr_v_t->GetYaxis()->SetTitle("v");
  gr_v_t->SetMarkerColor(kRed);
  gr_v_t->Draw("Psame");
```

```
c1->cd(2); // draw on pad 2
// v versus x
TGraph * gr_x_v = new TGraph(nsteps, &y0_v[0], &y1_v[0]);
gr_x_v->SetName("Graph_v_vs_x");
gr_x_v->GetXaxis()->SetTitle("x");
gr_x_v->GetYaxis()->SetTitle("v");
gr_x_v->SetTitle("Graph of v vs x");
gr_x_v->Draw("AP");
// Write the Canvas to the output ROOT file
c1->Write();
ofile->Close();

return s;
}
```

The code is relatively straightforward and consists of three functions:

- `int func(double t, ...)` implements the right-hand side of Eq. 2.24.
- `int jac(double t, ...)` populates the 2×2 Jacobian matrix (Eq. 2.25) using a `gsl_matrix` object passed via the `dfdy` argument.
- `int main(void)` orchestrates the main program execution, combining both GSL and ROOT functionality. GSL handles the numerical integration of the harmonic oscillator ODE system, while ROOT is used to visualize the results via `TGraph` and `TCanvas` objects, and to save the output using the `TFile` class.

Unlike the previous examples, the codes in this section are standalone and must be compiled and linked against the system's GSL and ROOT libraries. Assuming that both libraries are correctly installed and accessible from the command-line shell, the typical workflow on Linux or macOS is to write the code in an editor, save it as `gsl_harmosc.cc`, and compile it using the standard g++ compiler with a single command shown in code listing 2.15.

Code listing 2.15 Compiling the harmonic oscillator example code

```
$ g++ gsl_harmosc.cc -o gsl_harmosc `root-config --cflags --libs`
↪ `gsl-config --cflags --libs`
```

The output of this command is an executable file named `gsl_harmosc`, which can be run directly from the command line, as shown in Code listing 2.16. The terminal output displays the time, position, and velocity values at each step during the numerical evolution of the ODE system.

Code listing 2.16 Running the harmonic oscillator example executable

```
$ ./gsl_harmosc
1.00000e-02 9.99950e-01 -9.99983e-03
2.00000e-02 9.99800e-01 -1.99987e-02
3.00000e-02 9.99550e-01 -2.99955e-02
4.00000e-02 9.99200e-01 -3.99893e-02
...
```

An output file `out.root` is created that we can open in the terminal and plot the canvas with the results.

Code listing 2.17 Drawing the harmonic oscillator graphs using ROOT

```
$ root -l out.root
root [0]
Attaching file out.root as _file0...
(TFile *) 0x14e8352b0
root [1] c1->Draw()
```

The numerical solution of the harmonic oscillator ODE from the code is shown in Fig. 2.5, confirming the characteristic behaviour of harmonic motion.

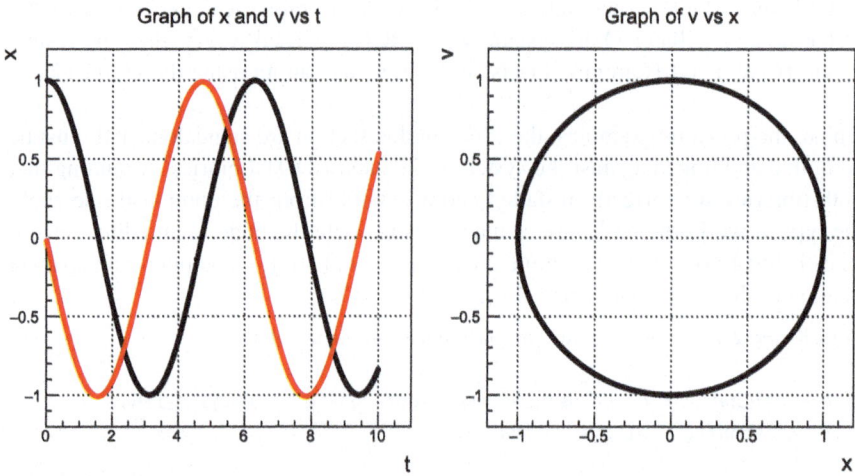

Fig. 2.5 Solution of the harmonic oscillator ODE system using a fixed-step fourth-order Runge-Kutta algorithm. Left: position x (black) and velocity v (red) as functions of time t. Right: phase-space trajectory showing position x versus velocity v

2.3.2 ODE Problem in the Complex Domain: Nuclear Magnetic Resonance

More complicated cases arise when functions and variables reside in the complex domain, as in the time-dependent Schrödinger equation.[7] This is particularly relevant for three-flavour neutrino propagation. In such problems, the differential equation involves complex-valued functions, though the evolution still occurs over a real-valued time variable. As a result, the same numerical algorithms used for real-valued ODEs can still be applied. However, the real and imaginary parts of the complex equations must be separated into distinct variables, leading to an extended system of real-valued ODEs—much like how we previously separated position and velocity components.

To prepare the necessary computational tools, we begin with a simpler, yet illustrative, example: nuclear magnetic resonance (NMR). NMR describes the time evolution of a two-state quantum system—typically a nuclear spin—where transitions between spin states are induced by an oscillating (radio-frequency) magnetic field.

Consider a spin-$\frac{1}{2}$ particle ($I = \frac{1}{2}$) initially placed in a static, external magnetic field $\mathbf{B} = (0, 0, B_z)$, oriented along the z-axis. We are free to align the z-axis of our coordinate system with the direction of this magnetic field. The interaction Hamiltonian is given by

$$H = -\mu\boldsymbol{\sigma} \cdot \mathbf{B}, \tag{2.26}$$

where μ is the magnetic moment of the particle, and $\boldsymbol{\sigma} = (\sigma_x, \sigma_y, \sigma_z)$ is the vector of the 2×2 Pauli matrices.

In this configuration, the particle undergoes precession around the axis defined by the magnetic field, meaning that the spin aligns with the field direction. This behaviour is naturally described in the basis associated with the Pauli matrices:

$$\sigma_x = \begin{pmatrix} 0 & 1 \\ 1 & 0 \end{pmatrix}, \ \sigma_y = \begin{pmatrix} 0 & -i \\ i & 0 \end{pmatrix}, \ \sigma_z = \begin{pmatrix} 1 & 0 \\ 0 & -1 \end{pmatrix}, \tag{2.27}$$

where only the σ_z has nonzero diagonal components. The time-dependent Schrödinger equation of the problem is

$$i\hbar\frac{d\Psi(t)}{dt} = -\mu\boldsymbol{\sigma} \cdot \mathbf{B}\Psi(t), \tag{2.28}$$

---

[7] Named after Austrian-Irish theoretical physicist E. Schrödinger (1887–1961).

where Ψ is a two-component state vector. Taking the basis states $|1\rangle \equiv |\uparrow\rangle \equiv \begin{pmatrix} 1 \\ 0 \end{pmatrix}$ and $|2\rangle \equiv |\downarrow\rangle \equiv \begin{pmatrix} 0 \\ 1 \end{pmatrix}$, a general state can be given as

$$|\Psi\rangle = c_1 \begin{pmatrix} 1 \\ 0 \end{pmatrix} + c_2 \begin{pmatrix} 0 \\ 1 \end{pmatrix} \equiv \begin{pmatrix} \psi_1 \\ \psi_2 \end{pmatrix}, \qquad (2.29)$$

where c_1 and c_2 are complex scalars normalized to one.

Due to the structure of the Pauli matrix basis, the two spin states, $|\downarrow\rangle$ and $|\uparrow\rangle$, correspond to alignment along the $-z$ and $+z$ directions, respectively. If the external magnetic field points only along the z-axis, the time evolution equations are completely decoupled, since the σ_z matrix has only diagonal elements. In this case, the two spin states have different energies: The state aligned parallel to the magnetic field has lower energy, while the antiparallel state has higher energy. The energy splitting between the two states depends on the strength of the magnetic field. When expressed in frequency units, the splitting is given by

$$\nu = \frac{\mu B_z}{hI}, \qquad (2.30)$$

where h is Planck's constant. For example, for a proton with magnetic moment $\mu_p \simeq 1.410606795 \times 10^{-26}$ J/T, a magnetic field of $B_z = 2.3488$ T leads to a transition frequency of approximately $\nu \simeq 200$ MHz.

Now consider a sample of protons (e.g. in water) at equilibrium in a constant magnetic field. There is a net magnetic moment equal to the vector sum of the individual moments aligned along the direction of **B**. If an oscillating magnetic field is briefly applied in a direction perpendicular to the static field, such that

$$\mathbf{B}(t) = \left(B_x \cos \omega_r t, \ B_y \sin \omega_r t, \ B_z \right), \qquad (2.31)$$

with oscillation frequency ω_r, then a *coupling* is introduced in the system's dynamics. This arises from the nonzero off-diagonal elements of the σ_x and σ_y Pauli matrices and appears explicitly in Eq. 2.28. As a result, transitions between the spin states ψ_1 and ψ_2 are induced, effectively flipping the spin between the $+z$ and $-z$ directions and releasing energy in the form of detectable radio-frequency signals.

We now express the problem in matrix-vector notation, leading to the following system of equations:

$$i\hbar \frac{d}{dt} \begin{pmatrix} \psi_1 \\ \psi_2 \end{pmatrix} = -\mu \left(\begin{pmatrix} 0 & 1 \\ 1 & 0 \end{pmatrix} B_x + \begin{pmatrix} 0 & -i \\ i & 0 \end{pmatrix} B_y + \begin{pmatrix} 1 & 0 \\ 0 & -1 \end{pmatrix} B_z \right) \begin{pmatrix} \psi_1 \\ \psi_2 \end{pmatrix}$$

$$= -\mu \begin{pmatrix} B_z \psi_1 + (B_x - iB_y)\psi_2 \\ (B_x + iB_y)\psi_1 - B_z \psi_2 \end{pmatrix}$$

Note that in the matrix form of the Schrödinger equation, the components of the wave function decouple if $B_x = B_y = 0$. However, when either B_x or B_y is nonzero, coupling arises between the two equations, and thus between the two spin states. This means that applying a magnetic field in a direction perpendicular to the constant field enables spin transitions or "flips". In fact, even a constant perpendicular component—such as a static B_x field—is sufficient to induce transitions.

Since the system is defined in the complex domain, both sides of the equations contain real and imaginary components. This results in a system of four real-valued first-order differential equations. The right-hand side of this system, which must be implemented in code, is evaluated using complex arithmetic. Fortunately, GSL provides built-in support for complex numbers, and the example shown in Code listing 2.18 demonstrates how to use it.

The initial condition of the wave function is set in the `main(void)` function as

```
double y[4] = {0.0, 0.0, 1.0, 0.0};
```

This corresponds to initializing the two-state system in one of the basis states, specifically:

$$\Psi_{\text{init}} = 0| \uparrow \rangle + 1| \downarrow \rangle = 0 \begin{pmatrix} 1 \\ 0 \end{pmatrix} + 1 \begin{pmatrix} 0 \\ 1 \end{pmatrix}. \tag{2.32}$$

In this simulation, we include both a constant magnetic field B_z and an oscillating B_x component. The time evolution is carried out using GSL's built-in Prince-Dormand Runge-Kutta method, a higher-order adaptive algorithm. It is worth noting that not all GSL solvers require a Jacobian matrix. Readers are encouraged to experiment with different algorithms to explore their numerical stability and performance.

Code listing 2.18 Numerical code for the evolution of the complex-domain NMR ODE system using GSL

```
#include <TFile.h>
#include <TGraph.h>
#include <TCanvas.h>
#include <TAxis.h>

#include <stdio.h>
#include <math.h>
#include <gsl/gsl_errno.h>
#include <gsl/gsl_odeiv2.h>
#include <gsl/gsl_complex.h>
#include <gsl/gsl_complex_math.h>
```

```
// Magnetic field values
double Bz = 0;
double Bx = 0;
double By = 0;
// Frequency
double wr = 0;
// Proton magnetic moment
const double mu_p = 1.410606795e-26; // J/T
// Planck's constant
const double hbar = 1.054571817e-34; // J*s

// Define the complex ODE system using gsl_complex
int complex_ode_func(double t, const double y[], double dydt[], void
↪   *params) {
    (void)(t); /* avoid unused parameter warning */

    gsl_complex psi_1 = gsl_complex_rect(y[0], y[1]);
    gsl_complex psi_2 = gsl_complex_rect(y[2], y[3]);

    // define the complex variable -i
    gsl_complex imag = gsl_complex_rect(0, -1);

    // the right-hand side of the diff. equation
    // using the GSL algebraic functions
    gsl_complex dpsi_1dt = gsl_complex_mul(imag, gsl_complex_add(
↪   gsl_complex_mul_real(psi_1, -Bz*mu_p/hbar),
↪   gsl_complex_mul_real(psi_2, Bx*(-mu_p/hbar)*cos(wr*t)) ));
    dpsi_1dt = gsl_complex_sub(dpsi_1dt, gsl_complex_mul_real(psi_2,
↪   By*(-mu_p/hbar)*sin(wr*t)));

    gsl_complex dpsi_2dt = gsl_complex_mul(imag, gsl_complex_sub(
↪   gsl_complex_mul_real(psi_1, -Bx*(mu_p/hbar)*cos(wr*t)),
↪   gsl_complex_mul_real(psi_2, -Bz*mu_p/hbar) ));
    dpsi_2dt = gsl_complex_add(dpsi_2dt, gsl_complex_mul_real(psi_1,
↪   By*(-mu_p/hbar)*sin(wr*t)));
    // Extract real and imaginary parts
    dydt[0] = GSL_REAL(dpsi_1dt);
    dydt[1] = GSL_IMAG(dpsi_1dt);
    dydt[2] = GSL_REAL(dpsi_2dt);
    dydt[3] = GSL_IMAG(dpsi_2dt);

    return GSL_SUCCESS;
}
```

```cpp
int main(void) {

  // Parameters of the problem
  Bz = 2.3488; // T
  Bx = Bz; // T
  By = 0; // T
  wr = 2*mu_p*Bz/hbar;  // Hz

  // Step size in s
  double h = 1e-07;
  // Define the GSL ODE system
  gsl_odeiv2_system sys = {complex_ode_func, NULL, 4, NULL};
  // The driver needs the system, stepping algo,
  // the step size and the abs, rel tolerances
  gsl_odeiv2_driver *d = gsl_odeiv2_driver_alloc_y_new(&sys,
↪   gsl_odeiv2_step_rk8pd, h, 1e-8, 0.0);
  double t = 0.0, t1 = 1e-04;
  // Initial value of the system:
  // RE(psi_1), IM(psi_1), RE(psi_2), IM(psi_2)
  double y[4] = {0.0, 0.0, 1.0, 0.0};
  // The probabilities
  double prob_psi2 = 0;
  double prob_psi1 = 0;

  // Save the intermediate step values
  std::vector<double> t_v, prob_psi1_v, prob_psi2_v;
  unsigned int nsteps = 0;

  // Stepping through the evolution of the ODE
  for (double ti = 0; ti <= t1; ti += 1e-07) {
    // Apply the ODE driver step
    int status = gsl_odeiv2_driver_apply(d, &t, ti, y);
    // Check for errors
    if (status != GSL_SUCCESS) {
      printf("error, return value=%d\n", status);
      break;
    }
    // Retrieve the current wave function values
    gsl_complex psi_2 = gsl_complex_rect(y[2],y[3]);
    gsl_complex psi_1 = gsl_complex_rect(y[0],y[1]);
    // Save the probabilities
    prob_psi2 = gsl_complex_abs2(psi_2);
    prob_psi1 = gsl_complex_abs2(psi_1);
```

```
  // Save the intermediate step values
  t_v.push_back(t);
  prob_psi2_v.push_back(prob_psi2);
  prob_psi1_v.push_back(prob_psi1);
  nsteps++;
}


gsl_odeiv2_driver_free(d);

// Create a ROOT output file to draw the result in graphs
// onto a canvas object
TFile * ofile = new TFile("out_nmr.root", "RECREATE");

// The canvas to draw the graphs
TCanvas *c = new TCanvas("c1", "c1",10,65,700,500);
c->Divide(2,1); // Divide to pads
// Draw to Pad 1
c->cd(1);
TGraph * gr_psi1_t = new TGraph(nsteps, &t_v[0], &prob_psi1_v[0]);
gr_psi1_t->Draw("APL");
gr_psi1_t->GetYaxis()->SetTitle("Probability of state |#uparrow>");
gr_psi1_t->GetXaxis()->SetTitle("t[s]");
gr_psi1_t->GetXaxis()->SetRangeUser(0, 1e-05);
// Draw to Pad 2
c->cd(2);
TGraph * gr_psi2_t = new TGraph(nsteps, &t_v[0], &prob_psi2_v[0]);
gr_psi2_t->Draw("APL");
gr_psi2_t->GetYaxis()->SetTitle("Probability of state |#downarrow>");
gr_psi2_t->GetXaxis()->SetTitle("t[s]");
gr_psi2_t->GetXaxis()->SetRangeUser(0, 1e-05);

// Write the Canvas to the output ROOT file
c->Write();
ofile->Close();

  return 0;
}
```

The code is compiled and executed in the same way as the harmonic oscillator example.

Code listing 2.19 Compiling and running the nuclear magnetic resonance example and plotting the results

```
$ g++ gsl_nmr.cc -o gsl_nmr `root-config --cflags --libs` `gsl-config
↪    --cflags --libs`
$ ./gsl_nmr
```

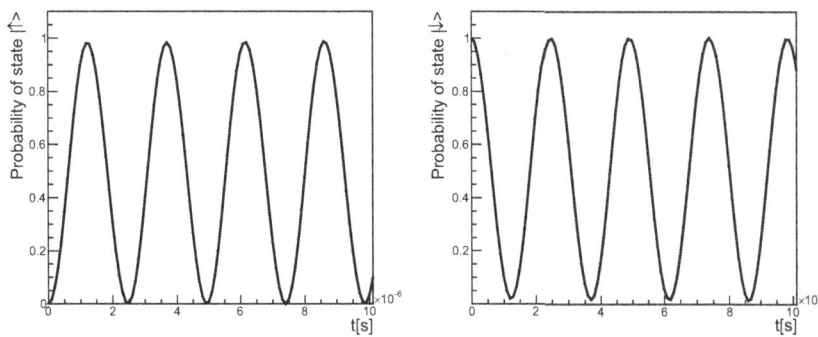

Fig. 2.6 Solution of the complex-valued NMR ODE system using GSL's Runge-Kutta Prince-Dormand algorithm. Left: probability of occupying the $| \uparrow \rangle$ state as a function of time t. Right: probability of occupying the $| \downarrow \rangle$ state over time

```
$ root -l out_nmr.root
root [0]
Attaching file out_nmr.root as _file0...
(TFile *) 0x13f7a4580
root [1] c1->Draw()
```

The result is shown in Fig. 2.6, where we set the oscillation frequency to $\omega_r = 2\mu_p B_z/\hbar \approx 628$ MHz. Over a time scale of 1 μs, the quantum state flips from the initial state $| \downarrow \rangle$ to $| \uparrow \rangle$ and then oscillates periodically between the two states.

2.4 Physical and Mathematical Constants

Numerical calculations often require values of fundamental constants, particle masses, and other physical properties. One way to access these values is through ROOT's built-in TParticlePDG and TMath classes, which allow users to query various properties of elementary particles and physical constants.

Particles in ROOT are identified by integer codes following the *Monte Carlo particle numbering scheme* established by the Particle Data Group (PDG). The full specification of this scheme is available at: https://pdg.lbl.gov/2020/reviews/rpp2020-rev-monte-carlo-numbering.pdf.

For example:

- The electron e^- (positron e^+) has a PDG ID of 11 (-11).
- The electron neutrino ν_e has PDG ID 12.
- The muon μ^- (antimuon μ^+) has PDG ID 13 (-13), and so on.

Nuclear PDG codes encode more information and follow a 10-digit format: $\pm$ 10LZZZAAAI. Here:

- L encodes the number of strange quarks (n_Λ).
- ZZZ is the total nuclear charge (the number of protons, n_p).
- AAA is the total baryon number: $n_p + n_n + n_\Lambda$.
- I indicates the isomer level (with $I = 0$ for the ground state).

For example, the most common stable isotope of iron is ^{56}Fe, which has $n_p = 26$, $n_n = 30$, $n_\Lambda = 0$, and is in the ground state ($I = 0$). Its corresponding PDG code is therefore

$$1000260560. \tag{2.33}$$

As of the time of writing (November 2023), the version of ROOT used in this manuscript (ROOT 6.28/06) includes only elementary particles in its internal database. Nonetheless, it remains useful for querying particle properties such as masses. The following example shows how to obtain the mass of the electron in GeV units:

Code listing 2.20 Obtaining the electron (PDGid 11) mass in an interactive ROOT session

```
root [0] TDatabasePDG * fPDGdb = new TDatabasePDG();
root [1] TParticlePDG * electron = fPDGdb->GetParticle(11);
root [2] electron->Mass()
(double) 0.00051099891
```

Several physical and mathematical constants can also be accessed via ROOT's TMath class. These are provided in SI units. For example:

Code listing 2.21 Querying the speed of light c and Planck's constant h in an interactive ROOT session

```
root [0] TMath::C()
(double) 2.9979246e+08
root [1] TMath::H()
(double) 6.6260701e-34
```

It is a good practice to cross-check the values of constants and particle properties from additional reputable sources. Two particularly useful resources are:

- The National Institute of Standards and Technology (NIST) Reference on Constants, Units, and Uncertainty: https://physics.nist.gov/cuu/Constants/index.html
- The Particle Data Group's Review of Particle Physics: https://pdg.lbl.gov/2023/tables/contents_tables.html

Table 2.3 Physical and mathematical constants used in the book

Physical and mathematical constants	
Quantity name, symbol	Value
Speed of light (vacuum), c	299,792,458 m/s
Reduced Planck constant, $\hbar$	$1.054571817 \times 10^{-34}$ J s
Reduced Planck constant times c	197.327 MeV fm
Electron charge magnitude, e	$1.602176634 \times 10^{-19}$ C
Electron volt, eV	$1.602176634 \times 10^{-19}$ J
Electron mass, m_e	0.510998950 MeV/c^2
Atomic mass unit, au	0.93149410242 GeV/c^2
Proton mass, m_p	1.007276466621 $\times$ au
Neutron mass, m_n	1.00866491595 $\times$ au
Deuteron mass, m_d	1.87561294257 GeV/c^2
Muon mass, m_μ	0.1056583755 GeV/c^2
Neutral Pion mass, m_{π^0}	0.1349768 GeV/c^2
Charged Pion mass, $m_{\pi^\pm}$	0.13957039 GeV/c^2
W-boson mass, m_W	80.377 GeV/c^2
W-boson vector coupling, c_V	1.0
W-boson axial-vector coupling, c_A	1.2754
Proton magnetic moment to nuclear magneton, μ_p	2.79284734463
Neutron magnetic moment to nuclear magneton, μ_p	-1.91304276
Fermi coupling constant, G_F	1.1663788 GeV^{-2}
Cabibbo angle, θ_c	0.23 rad
π	3.141592653589793238
Euler's e	2.718281828459045235
Cross section	10^{-28} m^2

Note that these sources may provide values in different unit systems. When incorporating constants into numerical codes, unit conversions may be required to ensure consistency.

To simplify usage across multiple programs, it is helpful to collect frequently used constants in a dedicated header file, for example, Constants.h. You can store this file in the same directory as your code files and include it in any program with a single line at the top:

```
#include "Constants.h"
```

The values of the constants used in the codes are those listed in Table 2.3. The source of the values is the Particle Data Group publication [6].

References

1. R. Brun, F. Rademakers, ROOT—an object oriented data analysis framework. Nuc. Instrum. Methods A **389**, 81–86 (1997). https://root.cern.ch/, https://github.com/root-project/root

2. M. Galassi et al., *GNU Scientific Library Reference Manual*, 3rd edn. Network Theory (2009)
3. J. von Neumann, Various techniques used in connection with random digits. Monte Carlo methods. Nat. Bureau Standards **12**, 36–38 (1951)
4. L.M. Chihara, T.C. Hesterberg, *Mathematical Statistics with Resampling and R*, 3rd edn. (Wiley, London, 2022)
5. T.C. Hesterberg, What teachers should know about the bootstrap: resampling in the undergraduate statistics curriculum. Am. Stat. **69**(4), 371–386
6. R.L. Workman, et al. (Particle Data Group), Prog. Theor. Exp. Phys **2022**, 083C01 (2022) and 2023 update

Relativistic Kinematics

3

Abstract

In this chapter, we explore key concepts and practical tools in relativistic kinematics. Topics include the transformation of four-vector between different reference frames, and the generation of initial- and final-state phase-space kinematics for decay and scattering events. We provide concrete examples demonstrating how to perform Lorentz boosts with ROOT's built-in classes. A discussion is devoted to the two-body and three-body phase-space factors, including illustrative examples involving two-body scattering and three-body decay processes.

3.1 Introduction

Perhaps the most important aspects of special relativity[1] are the existence of invariant quantities and conservation laws. These allow us to perform two essential types of computations: (i) relate physical quantities between different reference frames and (ii) determine which initial and final states are kinematically allowed in scattering and decay processes. In this chapter, we will not delve into the full formalism of relativity theory, but rather focus on the practical subset of rules needed for particle physics applications.

Assume we have two inertial frames, K and K', moving at constant velocity with respect to each other. Consider two events occurring at space-time coordinates $R_1 = (t_1, x_1, y_1, z_1)$ and $R_2 = (t_2, x_2, y_2, z_2)$. The experimentally established fact that the speed of light in vacuum is a universal constant—$c = 299{,}792{,}458$ m/s—in all inertial frames [1], allows us to express the spatial distance d between the two events in two equivalent ways:

---

[1] Developed by German-born theoretical physicist A. Einstein (1879–1955).

© The Author(s), under exclusive license to Springer Nature Switzerland AG 2026
B. Radics, *Neutrino Physics*, Lecture Notes in Physics 1043,
https://doi.org/10.1007/978-3-032-03993-4_3

43

$$d = \sqrt{(x_1 - x_2)^2 + (y_1 - y_2)^2 + (z_1 - z_2)^2}, \tag{3.1}$$

$$d = c(t_1 - t_2).$$

Since these expressions must describe the same physical distance, we define the differences $\Delta x = x_1 - x_2$ (and similarly for y, z, and t) and introduce the space-time interval:

$$\Delta s^2 = c^2 \Delta t^2 - \Delta x^2 - \Delta y^2 - \Delta z^2. \tag{3.2}$$

The quantity Δs^2 is an invariant under Lorentz[2] transformations—it has the same value in all inertial frames. In particular, if the same events are observed in the primed frame K', we must also have

$$\Delta s^2 = c^2 \Delta t'^2 - \Delta x'^2 - \Delta y'^2 - \Delta z'^2. \tag{3.3}$$

Let us look at a physical application: the lifetime of a decaying particle. Let $\Delta t = t_1 - t_0$ represent the time between the particle's creation at t_0 and its decay at t_1. We consider two frames: the rest frame K' (in which the particle is at rest) and the laboratory frame K (in which the particle moves with constant velocity v). In the rest frame, the particle does not change position, so $\Delta x' = \Delta y' = \Delta z' = 0$, and the interval simplifies to $\Delta s^2 = c^2 \Delta t'^2$, where $\Delta t'$ is the proper lifetime. In the laboratory frame, the particle travels a distance before decaying, so

$$\Delta s^2 = c^2 \Delta t^2 - \Delta x^2 - \Delta y^2 - \Delta z^2. \tag{3.4}$$

Equating the two expressions for Δs^2 gives

$$c^2 \Delta t'^2 = c^2 \Delta t^2 - \Delta x^2 - \Delta y^2 - \Delta z^2, \tag{3.5}$$

$$\Delta t'^2 = \Delta t^2 \left(1 - \frac{\Delta x^2 + \Delta y^2 + \Delta z^2}{c^2 \Delta t^2} \right), \tag{3.6}$$

$$\Delta t' = \Delta t \sqrt{1 - \frac{\Delta x^2 + \Delta y^2 + \Delta z^2}{c^2 \Delta t^2}} = \Delta t \sqrt{1 - \beta^2} = \frac{\Delta t}{\gamma}. \tag{3.7}$$

In the final step, we introduce the dimensionless velocity parameter $\beta \equiv v/c$ and the Lorentz factor $\gamma \equiv \frac{1}{\sqrt{1-\beta^2}}$, the reciprocal of which appears in the last expression. We obtained the well-known time dilation effect in special relativity. We can use the Taylor expansion of the square root function around $x = 0$ to recover the classical limit:

---

[2] Named after Dutch theoretical physicist H. Lorentz (1853–1928).

$$\sqrt{1-x} = 1 - \frac{1}{2}x - \frac{1}{8}x^2 - \frac{1}{16}x^3 - \dots \tag{3.8}$$

Applying this to the expression for the time dilation factor:

$$\Delta t' = \Delta t \sqrt{1-\beta^2} = \Delta t \left(1 - \frac{1}{2}\beta^2 - \frac{1}{8}\beta^4 - \dots \right). \tag{3.9}$$

In the limit of small velocities, i.e. $\beta \ll 1$, higher-order terms become negligible, and we recover

$$\Delta t' \approx \Delta t \left(1 - \frac{1}{2}\beta^2\right) \approx \Delta t. \tag{3.10}$$

Thus, in the low-speed limit, the proper time $\Delta t'$ approaches the coordinate time Δt, as expected from classical (nonrelativistic) physics; see Fig. 3.1. The two expressions agree closely for small values of $\beta \ll 1$, but diverge as β approaches 1, illustrating the importance of special relativity in particles physics.

To summarize, we have identified a linear transformation between time coordinates in two inertial frames moving at constant relative velocity. This transformation preserves the space-time interval Δs^2 and relies on a special metric—namely, the flat Minkowski metric,[3] of special relativity, denoted as $(1, -1, -1, -1)$ in the coordinate basis (ct, x, y, z). Note that in other coordinate systems (e.g. spherical), the metric would differ. Without deriving the full Lorentz transformation here, we now turn to linear transformations in space-time that affect both spatial and temporal coordinates.

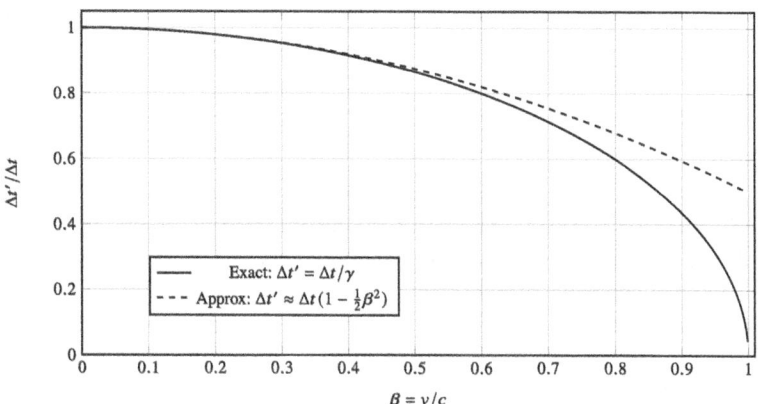

Fig. 3.1 Comparison of the exact time dilation formula with its Taylor approximation

---

[3] Named after mathematician H. Minkowski (1864–1909).

3.2   Lorentz Transformation

The behaviour of physical quantities—such as space-time coordinates, energy, mass, and momentum—under a change of inertial frame allows us to classify them according to how they transform.

We have already encountered an example of a *scalar* quantity: the space-time interval Δs^2, which remains unchanged under Lorentz transformations. We also introduced the concept of a *vector*, exemplified by the four-component space-time position vector $R = (t, x, y, z)$, which transforms according to a specific rule between inertial frames.

More generally, there are also *tensors*—such as the electromagnetic field strength tensor—which play an essential role in relativistic field theories but are beyond the scope of this book. Nonetheless, they are worth mentioning for completeness, and the interested reader is referred to standard texts (e.g. [2]).

In general, Lorentz transformations can be understood as rotations in space-time that preserve the space-time interval between events. Four-component quantities that transform in the same way as the four-position vector $R = (ct, x, y, z)$ are called *four-vectors*. The most common examples include four-position, four-momentum, and four-velocity. The general form of a linear transformation of a four-vector v can be written in matrix form as

$$v' = \Lambda v, \tag{3.11}$$

where Λ is a 4×4 matrix specific to the type of transformation being applied.

In this book, we are primarily concerned with Lorentz boosts, which correspond to changes between reference frames moving with constant relative velocity. For a boost along the x-axis, the transformation matrix (see derivation in [3]) is

$$\Lambda_x = \begin{pmatrix} \gamma & -\beta\gamma & 0 & 0 \\ -\beta\gamma & \gamma & 0 & 0 \\ 0 & 0 & 1 & 0 \\ 0 & 0 & 0 & 1 \end{pmatrix}, \tag{3.12}$$

where $\beta = v/c$ and $\gamma = 1/\sqrt{1 - \beta^2}$ are defined as before. Equation 3.12 shows that only the time and x components of the four-vector are affected by the boost; the y and z components remain unchanged.

A more general Lorentz boost in an arbitrary direction is parametrized by the velocity vector $\boldsymbol{\beta} = \mathbf{v}/c$, with magnitude $\beta = |\boldsymbol{\beta}|$. The corresponding transformation matrix is

$$\Lambda = \begin{pmatrix} \gamma & -\gamma\beta_x & -\gamma\beta_y & -\gamma\beta_z \\ -\gamma\beta_x & 1+(\gamma-1)\frac{\beta_x^2}{\beta^2} & (\gamma-1)\frac{\beta_x\beta_y}{\beta^2} & (\gamma-1)\frac{\beta_x\beta_z}{\beta^2} \\ -\gamma\beta_y & (\gamma-1)\frac{\beta_y\beta_x}{\beta^2} & 1+(\gamma-1)\frac{\beta_y^2}{\beta^2} & (\gamma-1)\frac{\beta_y\beta_z}{\beta^2} \\ -\gamma\beta_z & (\gamma-1)\frac{\beta_z\beta_x}{\beta^2} & (\gamma-1)\frac{\beta_z\beta_y}{\beta^2} & 1+(\gamma-1)\frac{\beta_z^2}{\beta^2} \end{pmatrix}. \tag{3.13}$$

When the boost is purely along the x-axis—that is, $\boldsymbol{\beta} = (\beta, 0, 0)$—Eq. 3.13 reduces to Eq. 3.12, as expected.

3.3   Energy-Momentum Four-Vector

Additional Lorentz-invariant quantities can be derived from four-vectors using the same flat space-time metric introduced earlier. In the previous subsection, we derived the transformation rule for the time coordinate of a particle between its rest frame and another frame in which it moves at constant velocity. This relationship motivates the definition of the *proper time*,

$$\tau \equiv \frac{t}{\gamma}, \tag{3.14}$$

where γ is the Lorentz factor. The proper time τ is the time measured in the particle's rest frame and is a fundamental scalar quantity in relativistic dynamics. Using proper time, we define the *four-velocity* η as

$$\eta = \frac{d}{d\tau}(ct, x, y, z) = \gamma\frac{d}{dt}(ct, x, y, z) = \gamma(c, \mathbf{v}). \tag{3.15}$$

To confirm that this four-vector has a Lorentz-invariant magnitude, we calculate a scalar quantity using the Minkowski metric (as in Eq. 3.2):

$$\eta^2 = \gamma^2 c^2 - \gamma^2 v^2 = \gamma^2 c^2(1 - \beta^2) = c^2.$$

Thus, the magnitude of the four-velocity is a constant, independent of the reference frame.

This four-velocity can now be used to define the *four-momentum* of a particle with rest mass m:

$$p \equiv m\eta = \gamma(mc, m\mathbf{v}) = \left(\frac{E}{c}, \mathbf{p}\right), \tag{3.16}$$

where $E = \gamma mc^2$ is the relativistic energy and $\mathbf{p} = \gamma m\mathbf{v}$ is the relativistic momentum of the particle. The invariant scalar corresponding to the four-momentum is

$$p^2 = m^2\eta^2 = m^2c^2, \tag{3.17}$$

again yielding a constant value, composed only of intrinsic properties of the particle (its rest mass) and the fundamental constant c.

Because the four-momentum is a four-vector, it transforms under Lorentz boosts in the same way as the four-position. Let us now consider an explicit example of a Lorentz boost applied to the four-momentum. Suppose a particle is at rest in its own frame, with four-momentum

$$p = (mc, 0, 0, 0),$$

and moves with velocity v in the $-x$ direction as seen from a laboratory frame.

From the rest frame's perspective, the lab frame is moving in the $+x$ direction. To determine the particle's four-momentum in the lab frame, we apply a Lorentz boost along the x-axis:

$$p' = \Lambda p = \begin{pmatrix} \gamma & -\beta\gamma & 0 & 0 \\ -\beta\gamma & \gamma & 0 & 0 \\ 0 & 0 & 1 & 0 \\ 0 & 0 & 0 & 1 \end{pmatrix} \begin{pmatrix} mc \\ 0 \\ 0 \\ 0 \end{pmatrix} = \begin{pmatrix} \gamma mc \\ -\beta\gamma mc \\ 0 \\ 0 \end{pmatrix}. \tag{3.18}$$

This result shows that, as measured in the lab frame, the particle has energy $E = \gamma mc^2$ and momentum $p_x = -\beta\gamma mc$. These expressions are consistent with the relativistic definitions of energy and momentum.

3.4   Experiments and Event Generators

In particle physics experiments, accelerators are used to collide particles in a well-defined *initial state* (beam and target), and a set of detectors captures the traces of particles in a random *final state* arising from various scattering or decay processes. Some of the fundamental questions we aim to answer include: How many total events should we expect after running collisions for a given time? What is the decay rate of a particle? How can theoretical models be compared with experimental data?

The total number of events, N, in an experiment is typically given by the *cross section*, σ, multiplied by the total beam *luminosity*, $\mathcal{L}$, and, in the case of a fixed-target experiment, the number of target atoms, $\mathcal{N}_t$:

$$N = \sigma \mathcal{L} \mathcal{N}_t. \tag{3.19}$$

The units are consistent: The cross section has a unit of area, L^2; the instantaneous beam luminosity (number of particles flowing through a unit area over unit time) with a unit of per area per time $L^{-2}T^{-1}$ is integrated over time to yield the total luminosity, with unit of per area L^{-2}; and $\mathcal{N}_t$ is dimensionless.

To describe particle decay, the *decay constant*, Γ, appears in the rate equation:

$$\frac{dN(t)}{dt} = -\Gamma N(t), \tag{3.20}$$

with the solution:

$$N(t) = N(0)e^{-\Gamma t}, \tag{3.21}$$

where $N(0)$ is the number of particles at $t = 0$, and $N(t)$ is the number remaining after time t. The *mean lifetime* is defined as $\tau = 1/\Gamma$, while the *half-life* is given by $t_{1/2} = \tau \ln 2$.

The decay constant Γ is often called the *decay width* because it is related to its mass distribution. In SI units, it has dimensions of inverse time, and in natural units, it is expressed in energy units. If a particle can decay via multiple channels, each channel i has its own partial decay width Γ_i. The total decay width is the sum of all partial widths:

$$\Gamma_{\text{tot}} = \sum_i \Gamma_i, \quad \text{or more generally} \quad \Gamma_{\text{tot}} = \int d\Gamma. \tag{3.22}$$

Some decay channels occur more frequently in certain ranges of solid angle or energy, denoted $[\Omega, \Omega+d\Omega]$ or $[E, E+dE]$. In such cases, we define the *differential decay rate*:

$$\Gamma_{\text{tot}} = \int \frac{d\Gamma}{d\Omega} d\Omega. \tag{3.23}$$

We are often not only interested in the total number of scattering events, N, but also in the distribution of reconstructed final-state particle kinematics—that is, the final-state *phase space*. Phase space refers to the set of generalized momentum states (as opposed to configuration space, which describes generalized coordinates). Understanding the shape of final-state distributions allows us to extract much more information when comparing theory and experiment.

The central theoretical quantity that describes the kinematic and dynamic characteristics of scattering is the *differential cross section*, $d\sigma/d\Omega$ or even $d^2\sigma/d\Omega dx$, where x can be any additional kinematic variable. Similarly, for decays, one works with the decay width $d\Gamma/d\Omega$. These quantities are usually provided by theory in functional form. Integrating the differential cross section yields the total cross section:

$$\sigma = \int \frac{d\sigma}{d\Omega} d\Omega. \tag{3.24}$$

In an experiment, the number of particles dN scattered into a solid angle element $d\Omega$ is given by

$$dN = \mathcal{L}\mathcal{N}_t \frac{d\sigma}{d\Omega} d\Omega, \tag{3.25}$$

from which we can extract the differential cross section:

$$\frac{d\sigma}{d\Omega} = \frac{1}{\mathcal{L}\mathcal{N}_t} \frac{dN}{d\Omega}. \tag{3.26}$$

In practice, detectors collect individual final-state particles that scatter into small angular ranges, $d\Omega$. Thus, carrying out an experiment is equivalent to *randomly sampling* from the differential cross section or decay width. The experimental challenge is to identify which process occurred and whether its cross section differs significantly from theoretical predictions. To compare theory with data, one must calculate the cross sections of all relevant processes, make predictions for their contributions in different regions of phase space, and assess the agreement between predictions and observations.

In summary, observing the final state of a single scattering or decay event amounts to a random sampling from either the differential cross section or the decay width. By accumulating many such events, we can map the magnitude and shape of these distributions. Furthermore, we can simulate experiments computationally by randomly sampling these distributions using a theoretical model. Such simulations, which generate synthetic events consistent with a given theoretical model, are known as *event generators*.

While in real experiments this random sampling is exact—nature produces final states according to physical laws—a computer code does not automatically follow these rules. When generating events in software, we must explicitly enforce physical constraints. Specifically, we must conserve energy and momentum and *weight* each process according to its differential cross section or decay width for the given kinematics. The following steps outline the general procedure for generating events in a computer algorithm for a scattering process $p_1 + p_2 \rightarrow p_3 + p_4 + \ldots$:

Event Generation in Computer Code
1. Define the *initial-state* (beam and target) four-momenta, p_1 and p_2, in a given reference frame.
2. Randomly generate four-momenta for the *final-state* particles, $p_3, p_4, \ldots$.
3. Ensure that *energy and momentum are conserved* between initial and final states.

(continued)

4. Use rejection sampling: Evaluate the function $d\sigma(p_1, p_2, \ldots)/d\Omega$ (or $d\Gamma/d\Omega$) for the given configuration, and determine whether the event is accepted. If accepted, this event might carry *event weights*.
5. Apply appropriate *Lorentz boosts* between relevant reference frames (e.g. Laboratory, Centre of Momentum, Nucleon at Rest) to obtain the final-state kinematics.

In both scattering and decay processes, phase-space factors appear naturally. The general form of the differential rate is

$$\text{Decay/scattering rate} \propto |\mathcal{M}|^2 \times R_n, \qquad (3.27)$$

where $\mathcal{M}$ is the quantum field theoretical matrix element (typically obtained from Feynman[4] rules), encapsulating the *interactions* between the particles, and R_n is the n-body final-state phase-space factor, which encodes kinematical weights and physical constraints.

We now proceed to study the structure of n-body phase-space factors in more detail.

3.4.1 Phase-Space Factors

The general expression for the Lorentz-invariant phase-space factor associated with a process going from an initial state $P_i = p_1 + p_2$ to a final state $P_f = p'_1 + p'_2 + \cdots + p'_n$ is given by

$$
\begin{aligned}
R_n = \int d^4 p'_1 \int d^4 p'_2 \ldots \int d^4 p'_n \\
\times \delta^4(p'_1 + p'_2 + \cdots + p'_n - p_1 - p_2) \\
\times \prod_{i=1}^{n} \delta(p'^2_i - m^2_i).
\end{aligned}
\qquad (3.28)
$$

This expression represents an integral over all possible final-state phase-space configurations. The four-dimensional Dirac δ-function[5] enforces conservation of energy and momentum. The additional product of δ-functions imposes the mass-shell condition for each final-state particle. Thus, the phase-space factor R_n acts

---

[4] Named after American theoretical physicist R. P. Feynman (1918–1988).
[5] Named after mathematician and theoretical physicist P. A. M. Dirac (1902–1984).

as a kinematic filter, assigning nonzero weights only to kinematically allowed final states and vanishing otherwise. Even among kinematically allowed configurations, the weighting is generally nonuniform across phase space.

Two well-known references in the literature [3, 5] provide a general recurrence formula for computing n-body Lorentz-invariant phase-space factors, well-suited for numerical implementation. Additionally, n-body phase-space factors can also be derived directly from the definition in Eq. 3.28; detailed derivation is included in Appendix A. These expressions are typically derived in the rest frame of the decaying particle but can be generalized to initial states with arbitrary four-momenta $P = p_1 + p_2$ by boosting into the centre-of-momentum (CM) frame.

The core idea is to reduce an n-body phase-space integral to a sequence of $(n-1)$-body integrals. For example, a three-body decay process $M \rightarrow m_1 + m_2 + m_3$ can be decomposed into two sequential two-body decays:

$$M \rightarrow M_{12} + m_3, \quad \text{followed by} \quad M_{12} \rightarrow m_1 + m_2,$$

where M_{12} is the two-particle system made up of particles m_1 and m_2. The kinematically allowed range for M_{12} is

$$m_1 + m_2 \leq M_{12} \leq M - m_3, \tag{3.29}$$

which reflects the minimal and maximal energy available to the $m_1 + m_2$ subsystem.

This structure allows us to assign a weight to any kinematically consistent configuration via

$$w_n = R_{n-l+1}(P; M, m_{l+1}, \ldots, m_n) \times R_l(P_l; m_1, \ldots, m_l), \tag{3.30}$$

where $R_{n-l+1}(P; M, m_{l+1}, \ldots, m_n)$ is the phase-space factor when there are $n - l$ particles, $m_{l+1}, \ldots, m_n$, plus one extra particle with mass $M^2 = P_l^2$. In this situation, the extra particle P_l represents the remaining system of $m_1, \ldots, m_l$ particles as a single object. And $R_l(P_l; m_1, \ldots, m_l)$ is the phase-space factor describing that extra particle, P_l. The weight w_n is to be normalized by the maximum weight, which will be given below.

The two-body phase-space factor for a process $P \rightarrow p_1 + p_2$ in the rest frame of a particle M ($P = (M, \mathbf{0})$; $M^2 = P^2 = (p_1 + p_2)^2$) is as follows (see Appendix A.1):

$$R_2(M; m_1, m_2) = \frac{\pi}{2M^2} \sqrt{[M^2 - (m_1 + m_2)^2][M^2 - (m_1 - m_2)^2]}$$

$$= \frac{\pi}{M} p(M; m_1, m_2), \tag{3.31}$$

where p is the magnitude of the three-momentum of the daughter particles in the CM frame, which we derive in A.7:

$$p^2(M; m_1, m_2) = \frac{[M^2 - (m_1 + m_2)^2][M^2 - (m_1 - m_2)^2]}{4M^2}.$$

From the general equation Eq. 3.30 above, we can express the three-body phase-space factor for the decay $P \to p_1 + p_2 + p_3$ as a product of two factors: (1) a phase-space factor describing the decay $P \to P_{1,2} + p_3$ with $P_{1,2}$ representing a composite "particle" $P_{1,2} = p_1 + p_2$, and (2) a phase-space factor describing the "decay" of the composite "particle" $P_{1,2} \to p_1 + p_2$, with $M_{1,2} = \sqrt{P_{1,2}^2} = \sqrt{(p_1 + p_2)^2}$:

$$R_3(M; m_1, m_2, m_3) = \frac{\pi^2}{M M_{1,2}} p(M; M_{1,2}, m_3) \times p(M_{1,2}; m_1, m_2). \qquad (3.32)$$

The maximum possible value of the weight $w_3^{\max}$ corresponds to the kinematic configuration where the momenta p are maximized. This occurs when

$$p^{\max}(M; M_{12}, m_3) = p(M; m_1 + m_2, m_3), \qquad (3.33)$$

$$p^{\max}(M_{12}; m_1, m_2) = p(M - m_3; m_1, m_2). \qquad (3.34)$$

Thus, the normalized three-body phase-space weight becomes

$$\frac{w_3}{w_3^{\max}} = \frac{R_3(M; m_1, m_2, m_3)}{R_3^{\max}(M; m_1, m_2, m_3)} = \frac{p(M; M_{12}, m_3)\, p(M_{12}; m_1, m_2)}{p(M; m_1 + m_2, m_3)\, p(M - m_3; m_1, m_2)}. \tag{3.35}$$

Note that the normalization factor is a constant, depending only on the particle masses. In contrast, the numerator varies across phase space due to the dependence of M_{12} on the momenta of the randomly selected final-state configuration. This introduces a nonuniform distribution of weights across the allowed region of phase space.

Importantly, for a two-body decay, the phase-space factor is constant (in the rest frame), as shown in Eq. 3.31.

3.5 Examples

3.5.1 Muons in Cosmic Radiation

The lifetime of the muon is approximately $\tau_\mu \simeq 2.2$ μs. Cosmic muons are produced in the upper layers of the Earth's troposphere (at an altitude of $h \simeq 10$ km) via the decay of hadrons generated by high-energy cosmic rays. However, even if these muons were traveling at the speed of light, c, the time it would take to reach the surface is

$$\Delta t = \frac{h}{c} \simeq \frac{10 \times 10^3 \text{ m}}{3 \times 10^8 \text{ m/s}} \simeq 33 \text{ μs,} \qquad (3.36)$$

which is significantly longer than their mean lifetime τ_μ in the rest frame. This apparent contradiction—muons reaching Earth's surface in large numbers despite their short lifetime—is resolved by relativistic time dilation.

From the muon's point of view, the atmosphere appears contracted due to the Lorentz contraction of lengths in the lab frame. To quantify this, we can calculate the effective distance the muon observes while traversing the atmosphere, accounting for relativistic effects.

The following ROOT code calculates the contracted path length using a Lorentz boost. Save the code as `MyBoost.C` and ensure that the previously prepared `Constants.h` file is available in the same directory. Then run the script interactively in ROOT, as shown below.

Code listing 3.1   Cosmic muons

```
#include "Constants.h"

void MyBoost(){

    // Approximate lifetime of the muon
    double dtMuon = 2.2e-06; // muon lifetime [s]

    // Kinematical variables
    double EMuon = 10.0; // muon energy [GeV]
    double h = 10000; // height [m]
    double gamma = EMuon/mMuon; // relativistic gamma
    double beta = TMath::Sqrt(1 - (1/(gamma*gamma)));
    double vMuon = beta*c; // muon velocity [m/s]


    // Print the relativistic factors
    cout << "beta: " << beta << ", gamma: " << gamma << ", beta*gamma: " <<
    ↪   beta*gamma << endl;


    // TLorrentzVector: 4-momentum vector of the muon in the Lab frame (px,
    ↪   py, pz, E)
    TLorentzVector p_Lab(TMath::Sqrt(EMuon*EMuon - mMuon*mMuon), 0, 0,
    ↪   EMuon);

    // Print the 4-momentum vector of the muon in the Lab frame
    cout << "\nThe 4-momentum vector of the muon in the Lab frame: " <<
    ↪   endl;
```

```
  p_Lab.Print();

  // ROOT's Lorentz Boost vector: b(px/E, py/E, pz/E)
  TVector3 beta_rest_to_lab = p_Lab.BoostVector();
  cout << "\nBoost vector: b(px/E, py/E, pz/E) " << endl;
  beta_rest_to_lab.Print();

  // Perform a Lorentz Boost back to the rest frame along the x-direction
  p_Lab.Boost( -beta_rest_to_lab);

  // Print the 4-momentum vector of the muon in it's rest frame
  cout << "\nThe 4-momentum vector of the muon in the Rest frame: " <<
↪    endl;
  p_Lab.Print();

  // In the Moun's rest frame its position (x,y,z,c*t) never changes but
↪    some time passes between it's birth and decay
  cout << "\nThe 4-position vector of the muon in the Rest frame: " <<
↪    endl;
  TLorentzVector R(0, 0, 0, c*dtMuon);
  R.Print();

  // Perform a Lorentz Boost to the Lab frame along the x-direction
  R.Boost( beta_rest_to_lab); // muon at rest to lab frame

  cout << "\nThe 4-position vector of the muon in the Lab frame: " <<
↪    endl;
  R.Print();
}
```

Code listing 3.2 Cosmic muons

```
root [0] .x MyBoost.C
beta: 0.999944, gamma: 94.6447, beta*gamma: 94.6394

The 4-momentum vector of the muon in the Lab frame:
(x,y,z,t)=(9.999442,0.000000,0.000000,10.000000)
↪    (P,eta,phi,E)=(9.999442,-0.000000,0.000000,10.000000)

Boost vector: b(px/E, py/E, pz/E)
TVector3 A 3D physics vector (x,y,z)=(0.999944,0.000000,0.000000)
↪    (rho,theta,phi)=(0.999944,90.000000,0.000000)

The 4-momentum vector of the muon in the Rest frame:
```

```
(x,y,z,t)=(-0.000000,0.000000,0.000000,0.105658)
 ↪   (P,eta,phi,E)=(0.000000,-0.000000,3.141593,0.105658)

The 4-position vector of the muon in the Rest frame:
(x,y,z,t)=(0.000000,0.000000,0.000000,659.543408)
 ↪   (P,eta,phi,E)=(0.000000,0.000000,0.000000,659.543408)

The 4-position vector of the muon in the Lab frame:
(x,y,z,t)=(62418.770757,0.000000,0.000000,62422.255169)
 ↪   (P,eta,phi,E)=(62418.770757,-0.000000,0.000000,62422.255169)
```

In Code Listing 3.1, we first define the relevant kinematic quantities for the case of a cosmic muon, such as its lifetime, rest mass, energy, and the altitude at which it is produced. From these, we calculate the dimensionless relativistic parameters β and γ, which we then use to perform the Lorentz boost transformation.

In ROOT, we can explicitly assign components to the four-momentum vector of a particle and keep track of the inertial frame before performing a Lorentz boost. We begin by setting up the four-momentum vector in the laboratory frame using the variable p_Lab, choosing the muon's motion to be along the x-axis. Note that the TLorentzVector class uses the argument ordering (p_x, p_y, p_z, E), meaning that the spatial momentum components come first.

The p_Lab.BoostVector() utility function provides a convenient way to compute the three-vector representation of the boost transformation, which is simply $b = (p_x/E, p_y/E, p_z/E)$. This follows from the relation $\beta_x = v_x/c = (mv_x)/(mc) = (\gamma m v_x)/(\gamma mc) = c\, p_x/E$, so the ratio p_x/E fully captures the needed boost factor. Since c is an invariant, it suffices to know $p_{x,y,z}/E$ and γ to compute the transformation.

With the obtained boost vector, we then perform a Lorentz boost back to the muon's rest frame using p_Lab.Boost(-beta_rest_to_lab). To confirm the result, we print the four-momentum vector after the boost. As expected, it shows that the spatial components vanish and the energy component equals the muon's rest mass: $(x, y, z, t) = (0, 0, 0, m_\mu)$. Note that the TLorentzVector class uses (x, y, z, t) as a generic label for four-vector components, so we must keep in mind that in this case it represents four-momentum: (p_x, p_y, p_z, E).

We now turn to the four-position vector. Since in the muon's rest frame it does not change its spatial position during its lifetime, the position four-vector is given as $R = (0, 0, 0, c\tau_\mu)$. We define this in the code using the variable R. When we boost this four-vector into the lab frame with the call R.Boost(beta_rest_to_lab), we obtain its components as observed in the lab. As expected, the time component increases due to time dilation, and the spatial x-component becomes nonzero.

The resulting x-component, as printed by the program, is approximately 62,418 m, which agrees with the theoretical calculation from the relation

$$x = \beta \gamma c \tau_\mu. \tag{3.37}$$

This result resolves the earlier apparent contradiction: Although the muon's proper lifetime is only about 2.2 μs, its relativistic motion allows it to cover more than 60 km in the lab frame. Therefore, a muon with $E = 10\,\text{GeV}$ can easily traverse the 10 km thickness of the Earth's troposphere before decaying.

Clearly, this example could have been computed using a single formula, and one might consider using ROOT an overkill. However, as we will see later, when dealing with thousands or millions of events involving decays or scattering of particles moving in different reference frames, automating the computation of boosts, momenta, and kinematic quantities is not only helpful but essential.

3.5.2 Photons from π^0 Decay

What is the angle between two photons from a π^0 decay as observed in the laboratory frame? Consider the decay process $\pi^0 \to \gamma + \gamma$, see Fig. 3.2.

In the rest frame of the pion, the two photons must be emitted back to back to conserve momentum:

$$\sum p_{\text{init}}^{\text{rest}} = p(0, 0, 0, m_{\pi^0}),$$

$$\sum p_{\text{final}}^{\text{rest}} = p_{\gamma_1}(E, 0, 0, E) + p_{\gamma_2}(-E, 0, 0, E).$$

Each photon carries half the energy of the parent particle, so $E = m_{\pi^0}/2$. Since photons are massless, we use the relation $|\mathbf{p}| = E$ in natural units. We are free to choose a coordinate system in which the π^0 is moving along the z-axis in the laboratory frame. We orient the back-to-back photon momenta along the x-axis in the pion rest frame. As shown in the code output below, in the rest frame the angle between the two photons is 180° by construction. However, in the laboratory frame, where the pion has total energy $E = 1\,\text{GeV}$ and is moving along the z-axis, the angle becomes much smaller. This angle is obtained by applying a Lorentz boost from the rest frame to the lab frame, and the code gives a result of approximately 15.5°.

(a)   (b)

Fig. 3.2 $\pi^0 \to \gamma\gamma$ decay in rest frame (left) and lab frame (right). (**a**) Decay at rest. (**b**) Decay in lab frame

Code listing 3.3 Photons from π^0 decay

```
#include "Constants.h"

void Photons_Pi0_Decay(){

  double EPion = 1.0; // pion energy [GeV]

  // 4-momentum of the pion in the Lab frame
  TLorentzVector p_Lab(0, 0, TMath::Sqrt(EPion*EPion - mPi0*mPi0),
↪  EPion);

  // 4-momentum of the pion in its rest frame
  TLorentzVector p_Rest(0, 0, 0, mPi0);

  // Back-to-back photons in the pion rest frame
  TLorentzVector p_Ph1(mPi0/2.0, 0, 0, mPi0/2.0);
  TLorentzVector p_Ph2(-mPi0/2.0, 0, 0, mPi0/2.0);

  // Three-momentum vectors
  TVector3 p3_Ph1(p_Ph1.X(), p_Ph1.Y(), p_Ph1.Z());
  TVector3 p3_Ph2(p_Ph2.X(), p_Ph2.Y(), p_Ph2.Z());

  // Angle in rest frame
  cout << "Angle between Ph1 and Ph2 in the rest frame: " <<
↪  p3_Ph1.Angle(p3_Ph2)*180/TMath::Pi() << " degrees" << endl;

  // Boost vector from rest frame to lab frame
  TVector3 beta_rest_to_lab = p_Lab.BoostVector();

  // Boost photons to lab frame
  p_Ph1.Boost(beta_rest_to_lab);
  p_Ph2.Boost(beta_rest_to_lab);

  // Angle in lab frame
  TVector3 p3_Ph1_l(p_Ph1.X(), p_Ph1.Y(), p_Ph1.Z());
  TVector3 p3_Ph2_l(p_Ph2.X(), p_Ph2.Y(), p_Ph2.Z());
  cout << "Angle between Ph1 and Ph2 in the lab frame: " <<
↪  p3_Ph1_l.Angle(p3_Ph2_l)*180/TMath::Pi() << " degrees" << endl;
}
```

Code listing 3.4 Console output of the π^0 decay code (Listing 3.3)

```
root [0] .x Photons_Pi0_Decay.C
Angle between Ph1 and Ph2 in the rest frame: 180 degrees
Angle between Ph1 and Ph2 in the lab frame: 15.5146 degrees
```

Let us verify that this numerical result is consistent with a direct Lorentz transformation calculation. Recall that a Lorentz boost corresponds to a matrix-vector operation, as in Eqs. 3.12 and 3.13. Adapting this to our case,[6] we apply the boost matrix to p_{γ_1}:

$$
p'_{\gamma_1} = \Lambda p_{\gamma_1} =
\begin{pmatrix}
\gamma & 0 & 0 & -\beta\gamma \\
0 & 1 & 0 & 0 \\
0 & 0 & 1 & 0 \\
-\beta\gamma & 0 & 0 & \gamma
\end{pmatrix}
\begin{pmatrix}
m_{\pi^0}/2 \\
m_{\pi^0}/2 \\
0 \\
0
\end{pmatrix}
=
\begin{pmatrix}
\gamma m_{\pi^0}/2 \\
m_{\pi^0}/2 \\
0 \\
-\beta\gamma m_{\pi^0}/2
\end{pmatrix}
\tag{3.38}
$$

$$
p'_{\gamma_2} =
\begin{pmatrix}
\gamma m_{\pi^0}/2 \\
-m_{\pi^0}/2 \\
0 \\
-\beta\gamma m_{\pi^0}/2
\end{pmatrix}.
\tag{3.39}
$$

The angle α between the two three-vector $\mathbf{p}'_{\gamma_1} = (m_{\pi^0}/2, 0, -\beta\gamma m_{\pi^0}/2)$ and $\mathbf{p}'_{\gamma_2} = (-m_{\pi^0}/2, 0, -\beta\gamma m_{\pi^0}/2)$ is calculated from their scalar product:

$$
\alpha = \cos^{-1}\left(\frac{\mathbf{p}'_{\gamma_1} \cdot \mathbf{p}'_{\gamma_2}}{|\mathbf{p}'_{\gamma_1}||\mathbf{p}'_{\gamma_2}|}\right) = \cos^{-1}\left(\frac{\beta^2\gamma^2 - 1}{\beta^2\gamma^2 + 1}\right)
$$

$$
= \cos^{-1}\left(\frac{(E^2/m_{\pi^0}^2) - 2}{(E^2/m_{\pi^0}^2)}\right) \simeq 0.2707801 \,\text{rad} = 15.514557°
$$

for $E = 1\,\text{GeV}$ and $m_{\pi^0} = 0.13497680\,\text{GeV}$.

Here we used the standard definitions $\gamma = E/m$ and $\beta = \sqrt{1 - 1/\gamma^2} = \sqrt{1 - m^2/E^2}$. This analytical result confirms the output of the code. It seems that we can trust the calculation performed by ROOT.

---

[6] Note that ROOT uses the convention where the time-like component of a four-vector is the last entry, whereas in the matrix equations shown here it appears first.

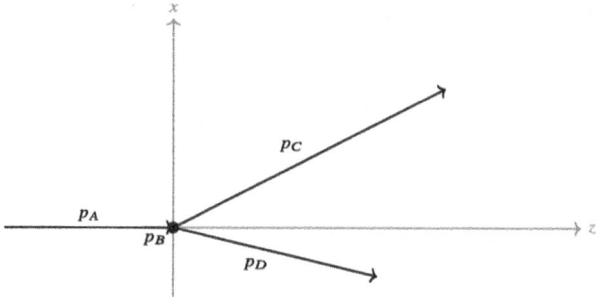

Fig. 3.3 Scattering in the laboratory frame: $p_A + p_B \rightarrow p_C + p_D$, where p_B is at rest

3.5.3 Neutrino Scattering on a Free Nucleon at Rest

We now explore how to sample the final-state phase space of the process $\nu + n \rightarrow p^+ + \mu^-$. As discussed earlier, sampling relativistic collision events in code begins with specifying the initial state and randomly selecting final states that satisfy energy and momentum conservation. This step alone does not yield physically accurate distributions—it merely selects *kinematically allowed* final states. In a later step, we must assign each event its proper physical weight in order to correctly model the actual distribution of final-state configurations. In this section, however, we focus purely on the kinematics.

Initial and final states are described in momentum space, using relativistic four-momentum notation: A four-vector is written as $p = (p_x, p_y, p_z, E) = (\mathbf{p}, E)$, and $p^2 = m^2$. For a generic $2 \rightarrow 2$ scattering process $p_A + p_B \rightarrow p_C + p_D$, energy-momentum conservation is encoded by the equation:

$$p_A + p_B = p_C + p_D. \tag{3.40}$$

The values of the four-momentum components depend on the chosen reference frame. In the case of a fixed-target experiment, where A is the incident particle (e.g. a neutrino) of mass m and three-momentum $\mathbf{p}$, and B is the stationary target of mass M (e.g. a neutron), we typically work in the laboratory frame, see Fig. 3.3. In this frame, the initial-state four-momenta are

$$p_A = (0, 0, |\mathbf{p}|, \sqrt{|\mathbf{p}|^2 + m^2}),$$

$$p_B = (0, 0, 0, M).$$

For ultrarelativistic particles such as neutrinos, the mass is negligible and we can approximate $|\mathbf{p}| \approx E$, so $p_A \simeq (0, 0, |\mathbf{p}|, |\mathbf{p}|)$.

Whatever method is used to generate random final states, we can always check whether the generated event is physically allowed by verifying whether the sum of the final-state four-momenta matches the initial-state total.

To demonstrate this in practice, let us write a short program using ROOT's TGenPhaseSpace class [3, 5]. This class provides a convenient way to randomly generate final states subject to energy-momentum conservation. After specifying the initial four-momentum and the final-state particle masses, calling the Generate() method produces a new random event—that is, a set of final-state four-vectors. These randomly generated final-state vectors are what we call *events*, just as in experimental particle collisions.

For this example, we simulate a beam of neutrinos with $|\mathbf{p}| = 1$ GeV (in natural units where $c = 1$) colliding with free neutrons at rest. The final state consists of a muon and a proton, approximating the kinematics of a charged-current quasi-elastic (CCQE) interaction. The code implementing this "experiment" is shown in code listing 3.5.

Code listing 3.5 Neutrino-nucleon scattering kinematics example

```
#include "Constants.h"

void Neutrino_nucleon_genps(){
  double mNu = 0.0; // approx. mass [GeV]

  // Initial state four-momenta
  double pBeam = 1.0; // momentum [GeV]
  TLorentzVector beam(0.0, 0.0, pBeam, pBeam);
  TLorentzVector target(0.0, 0.0, 0.0, mNeutron);

  // Form the beam particle
  TLorentzVector pInitial = beam + target;

  // Mass of the final-state particles
  double masses[2] = { mMuon, mProton};

  // Generate the decay in the nucleon at rest frame
  TGenPhaseSpace event;
  bool allowed = event.SetDecay(pInitial, 2, masses);

  for(unsigned int iEv = 0; iEv < 10; iEv++){
    Double_t weight = event.Generate();
    TLorentzVector *pMuon = event.GetDecay(0);
    TLorentzVector *pProton = event.GetDecay(1);
    cout << "--------------- E V E N T ---------------" << endl;
    cout << "Muon 4-mom: " << endl;
    pMuon->Print();
    cout << "Proton 4 -mom: " << endl;
    pProton->Print();
```

```
// Is energy-momentum conserved?
TLorentzVector pFinal = *pMuon + *pProton;
cout << "P initial: " << pInitial.Px() << ", " << pInitial.Py() << ",
  " << pInitial.Pz() << ", " << pInitial.E() << endl;
cout << "P final: " << pFinal.Px() << ", " << pFinal.Py() << ", " <<
  pFinal.Pz() << ", " << pFinal.E() << endl;
} // End of event loop
}
```

Note that in the code we can define the initial joint four-momentum vector as pInitial = beam + target, representing the sum of the two initial-state particles' four-momenta. This is particularly useful because we can always boost into the rest frame of this combined system and treat it as a particle that decays into an n-body final state.

Saving the code above in a file named Neutrino_nucleon_genps.C and executing it in a ROOT interactive session should yield an output similar to the following:

Code listing 3.6 Console output of the code listing 3.5

```
root [0] .x Neutrino_nucleon_genps.C
--------------- E V E N T ---------------
Muon 4-mom:
(x,y,z,t)=(-0.106570,-0.311531,-0.185520,0.392239)
Proton 4-mom:
(x,y,z,t)=(0.106570,0.311531,1.185520,1.547327)
P initial: 0, 0, 1, 1.93957
P final: 0, 0, 1, 1.93957
...
```

The output confirms that the total initial and final four-momenta are identical, thereby satisfying energy and momentum conservation.[7] This means that using TGenPhaseSpace at this stage of the event generation procedure ensures that the generated events are *kinematically allowed*.

However, the resulting final-state distributions are not yet physically accurate. As discussed earlier, the correct relative probabilities of different final-state configurations depend on the squared matrix element (amplitude) of the process, which is encoded in the differential cross section. We will delve into this important detail in the following chapters.

---

[7] You may also notice that the initial momentum in the transverse plane is zero, so the muon and proton emerge back to back in the transverse direction.

3.5.3.1 Two-Body Phase-Space Factors

As demonstrated in the previous code, TGenPhaseSpace generates so-called *weighted* events, where the event weight is computed from the n-body phase-space factor discussed earlier in the general integral expressions for decay widths or cross sections.[8] It is important to store the weight associated with each generated event and use these weights when plotting distributions. However, to interpret these weights meaningfully, we must verify what values TGenPhaseSpace actually produces.

This is straightforward to check using the explicit form of the two-body final-state phase-space factor from Eq. 3.31, which is

$$R_2(M; m_1, m_2) = \frac{\pi}{2M^2} \sqrt{[M^2 - (m_1 + m_2)^2][M^2 - (m_1 - m_2)^2]}. \qquad (3.41)$$

Substituting the values of the parent and daughter particle masses from code listing 3.5 yields a result of $R_2 \simeq 1.05777$. However, what we typically need for event generation is the *normalized weight*, which is the ratio of the phase-space weight to the maximum weight. In the case of two-body final states, this maximum value is simply the same as the expression above, since the value depends only on constants (masses) and is not a function of random kinematic variables.

For higher multiplicity final states (e.g. three-body, four-body, etc.), the situation is different: The phase-space weight depends on the specific configuration of final-state momenta (see Eq. 3.32), so the maximum must be computed numerically or from physical constraints. Nonetheless, many of the constant prefactors (such as π, or mass terms) cancel out when computing the normalized weights as ratios, see Eq. 3.35. Therefore, for practical purposes, we often work with a simplified form of the two-body phase-space weight (which also equals its maximum value):

$$R_2(M; m_1, m_2) = R_2^{\max} = \frac{1}{2M} \sqrt{[M^2 - (m_1 + m_2)^2][M^2 - (m_1 - m_2)^2]}. \qquad (3.42)$$

In this example, this yields a value of $R_2^{\max} \simeq 0.55956$. The function call event.GetWtMax() in TGenPhaseSpace returns the *inverse* of this value, namely $1/R_2^{\max} \simeq 1.78713$, which agrees with our calculation. This consistency can be verified by adding the following few lines of code to the previous example:

Code listing 3.7 Two-body phase-space factor example

```
// the masses of the particles
double a = pInitial.Mag(); // invariant mass of initial state
double b = mProton;        // mass of final-state proton
double c = mMuon;          // mass of final-state muon
```

---

[8] As we have seen before, the rejection sampling algorithm generated *unweighted* events.

```
// Get ROOT's maximal event weight
std::cout << "Wtmax: " << 1.0/event.GetWtMax() << std::endl;

// Check against our analytic calculation
std::cout << "2-body PS: "
          << (1.0/(2*a)) * TMath::Sqrt((a*a - (b+c)*(b+c)) * (a*a -
↪   (b-c)*(b-c)))
          << std::endl;
```

3.5.4   Decay of the Free Neutron at Rest

Neutron decay ($n \rightarrow p^+ + e^- + \bar{\nu}_e$) is a fundamental example that will be examined in greater detail in the next chapter. For now, we can use it to illustrate phase-space kinematics generation. Code listing 3.8 provides a simple implementation using TGenPhaseSpace.

In this case, there is no beam in the initial state—only a neutron at rest. The final state, however, involves three particles: a proton, an electron, and an electron antineutrino. In the laboratory frame, the initial four-momentum of the neutron is simply:

$$p = (0, 0, 0, M_n), \tag{3.43}$$

where M_n is the neutron rest mass. Since this is a three-body decay, the momenta of the final-state particles are not restricted to be back to back in the transverse plane.

Code listing 3.8 Neutron decay kinematics example

```
#include "Constants.h"

void Neutron_Decay_genps(){

  // Create a histogram to store the energy values
  TH1F * hE_El = new TH1F("hE_El", "Electron energy", 100, 0, 1.5);

  double mAntiNu = 0.0; // approx. mass

  // Initial state four-momentum
  TLorentzVector target(0.0, 0.0, 0.0, mNeutron);

  double masses[3] = { mProton, mElectron, mAntiNu};

  // Generate the Decay in the rest frame of the neutron
```

```
TGenPhaseSpace event;
bool allowed = event.SetDecay(target, 3, masses);

for(unsigned int iEv = 0; iEv < 10000; iEv++){
  Double_t weight = event.Generate();

  TLorentzVector *pProton = event.GetDecay(0);
  TLorentzVector *pElectron = event.GetDecay(1);
  TLorentzVector *pAntiNu = event.GetDecay(2);
  cout << "-------------- E V E N T --------------" << endl;
  cout << "Proton 4-mom: " << endl;
  pProton->Print();
  cout << "Electron 4-mom: " << endl;
  pElectron->Print();
  cout << "Antinu 4-mom: " << endl;
  pAntiNu->Print();

  // Fill histogram and convert to MeV
  hE_El->Fill(pElectron->E()*1000.0, weight);

  // Is energy-momentum conserved?
  TLorentzVector pFinal = *pProton + *pElectron + *pAntiNu;
  cout << "P initial: " << target.Px() << ", " << target.Py() << ", "
  << target.Pz() << ", " << target.E() << endl;
  cout << "P final: " << pFinal.Px() << ", " << pFinal.Py() << ", " <<
  pFinal.Pz() << ", " << pFinal.E() << endl;

} // End of event loop
// Plot the Electron energy distribution
hE_El->Draw("hist");
hE_El->GetXaxis()->SetTitle("E [MeV]");
hE_El->GetYaxis()->SetTitle("Events");
}
```

In this example, we use the TH1F class of ROOT to create a histogram of the final-state electrons' energy. We fill the histogram with values obtained from each generated event, using the corresponding event weights from TGenPhaseSpace. The resulting distribution is shown in Fig. 3.4.

The boundaries of the distribution are physically accurate: The minimum total energy of the electron corresponds to its rest mass, $m_e = 0.511$ MeV/c^2, while the endpoint reflects the maximal possible energy the electron can receive in the decay, which equals the mass difference between the neutron and the proton, $m_n - m_p \simeq 1.29$ MeV/c^2.

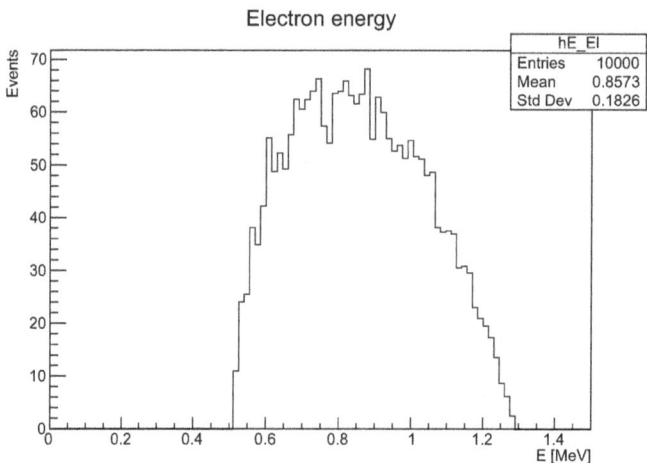

Fig. 3.4 Distribution of the final-state energy of electrons from the free neutron decay, using only kinematic phase-space weighting

If we compare this with the physically correct distribution (see the chapter on neutron decay in [4]), we observe that the overall shape of the generated kinematics is already a reasonable approximation of the physical shape. This is primarily because the small mass difference between the neutron and the proton significantly constrains the final-state phase space. In contrast, as we will see later, in the case of the muon decay, $\mu^- \rightarrow e^- + \bar{\nu}_e + \nu_\mu$, the energy spectrum of the final-state electrons is broader; the phase space opens up due to the large mass available for the decay: $m_\mu = 105.7 \text{ MeV/c}^2$, $m_e = 0.511 \text{ MeV/c}^2$. Nevertheless, the generated electron energy distribution from the neutron decay is not entirely physical yet, as it does not account for the dynamics of the decay process.

As we will see in the upcoming sections, a physically correct energy spectrum requires weighting each event not only by the kinematics (i.e. phase space), but also by the square of the matrix element. In practice, this is implemented by evaluating the differential decay width and using it as an effective weight. We will achieve this via the technique of rejection sampling.

The terminal output of the code confirms that energy and momentum are conserved between the initial and final states. The small deviations from zero in the transverse components (of order 10^{-20}) are numerical artefacts and can be safely interpreted as zero. An example output from a generated event is shown below.

Code listing 3.9 Example console output of the code listing 3.8

```
--------------- E V E N T ---------------
Proton 4-mom:
(x,y,z,t)=(0.000382,-0.000567,0.000610,0.938272)
Electron 4-mom:
```

```
(x,y,z,t)=(-0.000056,0.000085,-0.000125,0.000536)
Antinu 4-mom:
(x,y,z,t)=(-0.000326,0.000482,-0.000484,0.000757)
P initial: 0, 0, 0, 0.939565
P final: 0, -5.42101e-20, 5.42101e-20, 0.939565
...
```

3.5.4.1 Three-Body Phase-Space Factors

In analogy with the two-body phase-space calculation, we can also compute the three-body phase-space factor and verify that the maximal event weight returned by TGenPhaseSpace matches our expectations. Recall from Eqs. 3.32 and 3.33 that the three-body phase-space weight is not constant and depends on the momenta of the final-state particles. However, its maximum value is given by a purely algebraic expression involving only the particle masses.

We can evaluate the maximal three-body phase-space factor explicitly with the following expression:

$$R_3^{\max}(M; m_1, m_2, m_3) = p^{\max}(M; m_1 + m_2, m_3) \times p^{\max}(M - m_3; m_1, m_2)$$

$$(3.44)$$

$$= \frac{1}{4M(M - m_3)} \sqrt{[M^2 - ((m_1 + m_2) + m_3)^2][M^2 - ((m_1 + m_2) - m_3)^2]}$$
$$\times \sqrt{[(M - m_3)^2 - (m_1 + m_2)^2][(M - m_3)^2 - (m_1 - m_2)^2]}.$$

We implement this in code below in code listing 3.10 and compare the result against the inverse of the maximal weight obtained via event.GetWtMax() from the TGenPhaseSpace class.

Code listing 3.10 Three-body phase-space factor: evaluation of maximal weight

```
// The masses of the particles
double M = mNeutron;
double m1 = mAntiNu;
double m2 = mProton;
double m3 = mElectron;

// Get ROOT's maximal event weight
std::cout << "Wtmax: " << 1.0/event.GetWtMax() << std::endl;

// Check against our calculation
double p1 = TMath::Sqrt((M*M - (m1+m2+m3)*(m1+m2+m3))*(M*M -
↪   (m1+m2-m3)*(m1+m2-m3))) / (2*M);
```

```
double p2 = TMath::Sqrt(((M-m3)*(M-m3) - (m1+m2)*(m1+m2))*((M-m3)*(M-m3)
↪    - (m1-m2)*(m1-m2))) / (2*(M-m3));

std::cout << "3-body PS: " << p1 * p2 << std::endl;
```

The result, $R_3^{max} \simeq 9.285 \times 10^{-7}$, should match the inverse of the maximal event weight returned by `TGenPhaseSpace`, confirming the consistency of the implementation with the theoretical expression.

3.5.5 Suppression of Dalitz Decays for Neutral Pion

Sometimes there are processes in which QED plays a large role and allows one to get a rough estimate of the branching fractions. The neutral pion decays to two photons with a very high branching ratio, $B(\pi^0 \rightarrow \gamma\gamma) \approx 98.8\%$. The next most frequent channel is the Dalitz decay [6], $\pi^0 \rightarrow e^+e^-\gamma$, with a branching ratio of approximately 1.2% [1].

A reasonable first-order estimate of this suppression can be obtained by comparing the ratio of the powers of the fine-structure coupling constant, $a^3/a^2 = a \approx 0.007$, dominated by the requirement of an additional QED vertex. This suggests that the reduced branching ratio is not due to the more restrictive phase space available in the three-body final state. An example implementation of the ratio of phase-space-only factors is shown in code listing 3.11.

Code listing 3.11 Estimating the phase-space ratio

```
#include "TGenPhaseSpace.h"
#include <iostream>

#include "Constants.h"

void Pi0_Decay_BR() {

    TLorentzVector pi0(0, 0, 0, mPi0);

    // --- 2-body decay: pi0 -> gamma gamma ---
    double masses2[2] = {0, 0};
    TGenPhaseSpace event2;
    event2.SetDecay(pi0, 2, masses2);
    double w2 = event2.GetWtMax();
    double sumWeight2 = 0;
    for (int i = 0; i < 1e7; ++i)
      sumWeight2 += event2.Generate();
```

```
// --- 3-body decay: pi0 -> gamma e+ e- ---
double masses3[3] = {0, mElectron, mElectron};
TGenPhaseSpace event3;
event3.SetDecay(pi0, 3, masses3);
double w3 = event3.GetWtMax();
double sumWeight3 = 0;
for (int i = 0; i < 1e7; ++i)
  sumWeight3 += event3.Generate();

// --- Qualitative ratio from phase-space ratio ---
std::cout << "(pi0->gamma e+ e- / pi0->gamma gamma) = " << 100.0 *
↪  sumWeight3 / sumWeight2 << " %" << std::endl;

}
```

The console output for a typical execution is shown in listing 3.12. The phase-space-only estimated value differs wildly from the experimental result because QED effects have been neglected, but it nevertheless provides a kinematic shape factor, illustrating how narrower the phase-space available for the Dalitz decay.

Code listing 3.12 Example console output of the code listing 3.11

```
root [0] .x Pi0_Decay_BR.C
  (pi0->gamma e+ e- / pi0->gamma gamma) = 25.2765 %
```

References

1. R.L. Workman et al. (Particle Data Group), Prog. Theor. Exp. Phys. **2022**, 083C01 (2022) and 2023 update
2. S.M. Carroll, *Spacetime and Geometry: An Introduction to General Relativity* (Cambridge, 2019)
3. R. Hagedorn, *Relativistic Kinematics* (W. A. Benjamin, 1964)
4. D.J. Griffiths, *Introduction to Elementary Particles*, 2nd edn. (Wiley & Sons, London, 2004)
5. F. James, *Monte Carlo Phase Space*. https://cds.cern.ch/record/275743 (CERN 1968)
6. R. Dalitz, On an alternative decay process for the neutral π-meson. Proc. Phys. Soc. London Sect. A **64**, 667 (1951)

Weak Decays

4

Abstract

This chapter extends the methodology of event generation from simple phase-space sampling to incorporating differential quantities using the rejection sampling technique. Leading-order muon and neutron decay processes are employed to introduce rejection sampling and numerical integration for calculating particle lifetimes. The generation of three-body final-state phase space is described in detail. As a further application of phase-space sampling, parts of the Solar neutrino energy spectrum are estimated. Finally, the decay of charged pions is used to demonstrate how phase-space kinematics underlies the basic principle of accelerator-based neutrino beams.

4.1   Muon Decay

Many important processes in elementary particle physics are governed by weak interactions. In this section, we focus mostly on decays, although the lessons here will prove valuable when we study scattering processes later on. Weak interactions are mediated by the $W^{\pm}$ and Z^0 bosons, and, at leading order in perturbation theory, they can be efficiently simulated using numerical methods. We will not delve into the underlying quantum field theory here; for a graduate-level introduction, the reader is referred to standard textbooks such as [1].

In the previous chapter, we discussed how to generate phase-space kinematics for scattering and decay events. Here, we build upon that foundation by incorporating the correct dynamics using the leading-order differential decay width in weak perturbation theory. Recall that event generation using the class TGenPhaseSpace ensures only energy and momentum conservation. At this stage, the generator does not yet account for the nature of the underlying interaction. To correct for this, we

© The Author(s), under exclusive license to Springer Nature Switzerland AG 2026   71
B. Radics, *Neutrino Physics*, Lecture Notes in Physics 1043,
https://doi.org/10.1007/978-3-032-03993-4_4

apply event weights using the appropriate differential decay width and the rejection sampling algorithm.

As an example, the muon decays almost exclusively via

$$\mu^- \rightarrow \nu_\mu + \bar{\nu}_e + e^-,$$

as illustrated in Fig. 4.1. According to Fermi's golden rule, the differential decay rate at leading order is given by

$$d\Gamma_\mu = \frac{\langle|\mathcal{M}|^2\rangle}{2\hbar m_\mu} \left(\frac{c\,d^3\mathbf{p_2}}{(2\pi)^3 2E_2}\right) \left(\frac{c\,d^3\mathbf{p_3}}{(2\pi)^3 2E_3}\right) \left(\frac{c\,d^3\mathbf{p_4}}{(2\pi)^3 2E_4}\right)$$
$$\times (2\pi)^4 \delta^4(p_1 - p_2 - p_3 - p_4). \tag{4.1}$$

We have kept the constants c and $\hbar$ explicit for completeness, though in natural units we take $\hbar = c = 1$. Note that the squared matrix element $\langle|\mathcal{M}|^2\rangle$ is dimensionless. In general, its dimension is given by E^{4-n}, where n is the total number of incoming and outgoing legs in the Feynman diagram. Since decay widths must have dimensions of energy, we can check dimensional consistency: the $1/m_\mu$ prefactor contributes E^{-1}, the three $d^3\mathbf{p}$ terms give E^9, the three $2E$ denominators yield E^{-3}, and the four-dimensional Dirac-delta function contributes E^{-4} (recall the identity $\delta(kx) = \delta(x)/|k|$). Altogether, the dimension is $E^{9-(1+3+4)} = E^1$, as expected for a decay width (see Sect. 2.1).

The dynamics of the decay are encoded in the squared, spin-averaged amplitude,[1] which for both muon and neutron decays takes a similar leading-order form:

$$\langle|\mathcal{M}|^2\rangle = 2\left(\frac{g_W}{M_W c}\right)^4 (p_1 \cdot p_2)(p_3 \cdot p_4). \tag{4.2}$$

Here, $g_W \simeq 0.66$ is the weak fermion-boson coupling, $M_W \simeq 80$ GeV, and p_i are the four-momenta of the particles involved, as shown in Fig. 4.1. In practice, we often work with the Fermi coupling constant,

$$G_F = \frac{\sqrt{2}}{8}\left(\frac{g_W}{M_W}\right)^2 \simeq 1.166 \times 10^{-5} \text{ GeV}^{-2},$$

expressed in natural units.

The experimental value of the muon lifetime is $\tau_\mu \simeq 2.2 \times 10^{-6}$ seconds. The theoretical leading-order expression is obtained by integrating Eq. 4.1. We quote the result here and will later verify it via numerical integration in code:

---

[1] Averaged over initial and summed over final spin states.

Fig. 4.1 Leading-order
Feynman diagram of the
weak decay of the muon

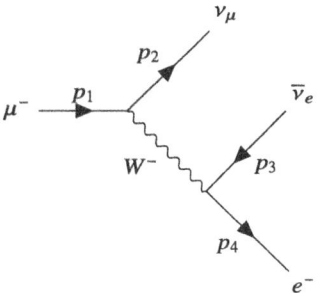

$$\tau_\mu = \frac{1}{\Gamma_\mu} = \frac{192\pi^3}{G_F^2 m_\mu^5}. \tag{4.3}$$

Substituting the numerical values for m_μ and G_F, we find $\tau_\mu \simeq 3.323 \times 10^{18}\,\text{GeV}^{-1}$ in natural units. Using the conversion factor from Table 2.1, we obtain the value in SI units:

$$\tau_\mu = 3.323 \times 10^{18}\,\text{GeV}^{-1} \times 6.583 \times 10^{-25} \rightarrow 2.2\,\mu\text{s},$$

in excellent agreement with the experimental result.

Let us now turn our attention to writing a simple event generator for this decay process. Specifically, we aim to obtain the energy distribution of the final-state electron from a muon decaying at rest. To do this, we need an expression for the differential decay width as a function of the electron energy. The standard result for the unpolarized case (using the spin-averaged matrix element $\langle |\mathcal{M}|^2 \rangle$) is

$$\frac{d\Gamma}{dE} = \left(\frac{8G_F}{\sqrt{2}}\right)^2 \frac{m_\mu^2 E^2}{2(4\pi)^3} \left(1 - \frac{4E}{3m_\mu}\right), \tag{4.4}$$

where E is the total energy of the electron.

To implement this in code, we proceed as before: We use the `TGenPhaseSpace` class to generate the kinematics for the decay of a muon at rest into three final-state particles. This ensures that energy and momentum are conserved. We then extend the generation by applying the rejection sampling algorithm (discussed in Sect. 2.2.1). The function in Eq. 4.4 is evaluated for each generated event, and the event is accepted or rejected accordingly.

We begin by generating the basic kinematics of the decay. See code listing 4.1. Save the code in a file named `MuDecay.C` and execute it in a ROOT interactive session using:

```
root [0] .x MuDecay.C
```

Code listing 4.1 Muon decay kinematics example

```cpp
#include <TGenPhaseSpace.h>
#include <TH1F.h>
#include <TRandom2.h>
#include <TLorentzVector.h>

#include <iostream>

#include "Constants.h"

void MuDecay(){

  // ROOT's random number generator
  TRandom2 * ran = new TRandom2(0);

  // Histogram to store the electron energies
  TH1F *hE_ElectronPSW = new TH1F("hE_ElectronPSW","", 60,0,60);
  TH1F *hE_Electron = new TH1F("hE_Electron","", 60,0,60);

  // Main event loop
  Int_t Nevents = 1e+05;
  for (Int_t nEv=0;nEv<Nevents;nEv++) {

    // initial-state muon at rest
    TLorentzVector pMuon(0.0,0.0, 0.0, mMuon); //GeV (beam "particle")

    // final-state particle masses
    Double_t masses[3] = { mElectron , mNu, mNu};

    // Generate the decay in then nucleon-at-rest frame
    TGenPhaseSpace event;
    bool allowed = event.SetDecay(pMuon, 3, masses);

    // 3-body Phase-space factor weight
    Double_t weight = event.Generate();

    // Obtain the four-momenta of the final-state particles
    TLorentzVector *pElectron = event.GetDecay(0);
    TLorentzVector *pNu1  = event.GetDecay(1);
    TLorentzVector *pNu2  = event.GetDecay(2);

    // Get the final-state electron energy
    double EEl = pElectron->E();
```

```
    // apply the Rejection Sampling algorithm - TODO -

    // Fill the electron energy with PS weight only
    hE_ElectronPSW->Fill(EE1*1000, weight); // GeV->MeV */
}// Done with event loop

    // Display the result
    hE_ElectronPSW->Draw("hist");
    hE_ElectronPSW->GetXaxis()->SetTitle("E [MeV]");
    hE_ElectronPSW->GetYaxis()->SetTitle("N. of events");
    hE_ElectronPSW->SetLineColor(kRed);
}
```

The result of the event generator is shown in Fig. 4.2. The red histogram displays the electron energy distribution when only phase-space weights are applied. Two physical boundaries are immediately apparent: (1) when the entire available energy is transferred to the electron, its energy reaches the maximum value, $E_e = m_\mu/2 \simeq$ 52.5 MeV; and (2) at threshold, the electron energy is equal to its rest mass, $E_e =$ 0.511 MeV, which lies in the first bin of the plot. However, not all intermediate configurations are equally likely: The phase-space weight depends on the momenta of the daughter particles.

The three-body phase-space distribution is visualized in Fig. 4.3, which shows the phase-space weights as a function of the energies of one of the neutrinos and the electron. This plot can be interpreted as follows. The points with the lowest weights correspond to extreme configurations near the physical boundaries, where

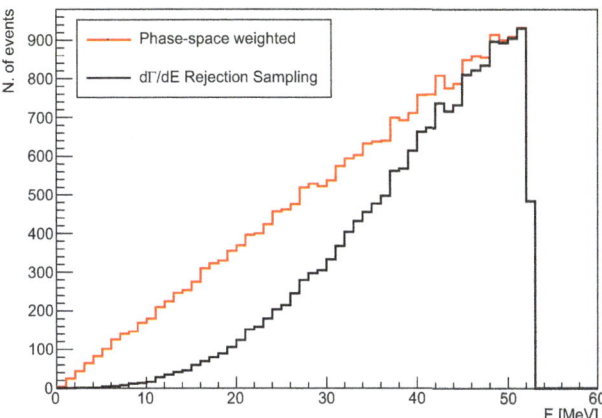

Fig. 4.2 Electron energy distribution from the $\mu \rightarrow \nu_\mu + \bar{\nu}_e + e$ muon decay process. The red histogram shows the distribution with phase-space weights only, and the black histogram shows the case of phase-space weights together with rejection sampling using the $d\Gamma/dE$ differential decay width function

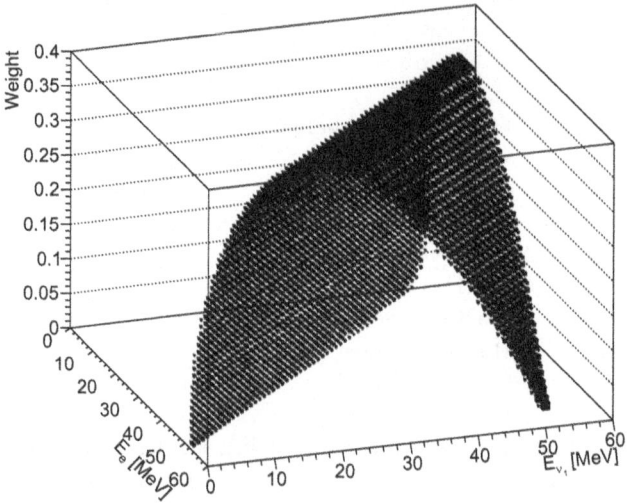

Fig. 4.3 Three-body phase-space weight distribution as a function of the energy of the electron one of the neutrinos

almost all the energy is carried by just two of the three final-state particles. For example: (i) $E_{\nu_1} \simeq 52.5$ MeV, $E_{\nu_2} \simeq 0$ MeV, $E_e \simeq 52.5$ MeV; (ii) $E_{\nu_1} \simeq 0$ MeV, $E_{\nu_2} \simeq 52.5$ MeV, $E_e \simeq 52.5$ MeV; or (iii) $E_{\nu_1} \simeq 52.5$ MeV, $E_{\nu_2} \simeq 52.5$ MeV, $E_e \simeq 0$ MeV. These lie near the kinematic edges of the allowed phase space and are statistically suppressed.

The diagonal feature at the base of the 3D plot corresponds to configurations satisfying $E_e + E_{\nu_1} = E_{\nu_2} \simeq 52.5$ MeV. This linear relationship arises from energy conservation in the muon rest frame, where the total energy of the final-state particles must equal the muon mass: $E_e + E_{\nu_1} + E_{\nu_2} = m_\mu$. Configurations near the boundary are limited in phase-space volume and therefore less probable. In contrast, balanced configurations where energy and momentum are more evenly distributed among the three particles occupy a larger region of phase space and have higher weights.

To illustrate this, consider the case where the electron energy is fixed at its maximum value, $E_e \simeq 52.5$ MeV. The remaining energy must be distributed between the two neutrinos. Although their vector momenta must sum to $-\mathbf{p}_e$ to conserve total momentum, the angle between the two neutrinos remains unconstrained, allowing many kinematically allowed combinations. Consequently, such configurations contribute significantly to the total weight.

On the other hand, when the electron energy is near zero, $E_e \simeq 0$ MeV, the two neutrinos must be emitted back to back to conserve momentum. This configuration is highly constrained, leaving essentially no freedom in momentum space, and thus carries very little weight. For intermediate electron energies, the number of possible momentum-sharing configurations increases with E_e. This explains why, in Fig. 4.2, the red histogram rises with electron energy: Low-energy electrons are rare, while high-energy electrons are much more probable due to the larger available phase space.

To obtain the physically correct shape of the electron energy distribution, the generated events must be weighted according to the differential decay width, $d\Gamma/dE_e$. This can be achieved using the familiar rejection sampling algorithm introduced earlier. In practice, this requires adding an additional code block to our existing event generator, inserted just below the following line, see code listing 4.2.

Code listing 4.2 Placeholder for implementing the rejection sampling part of the code

```
// apply the Rejection Sampling algorithm - TODO
```

A possible implementation is shown in code listing 4.3.

Code listing 4.3 Rejection sampling applied to the muon decay kinematics

```
// apply the Rejection Sampling algorithm - TODO

// Evaluate dGamma/dE
double r = dGamma_dE(EE1);

//----------------------------------------------------
// We apply the Rejection Sampling algorithm:
// Need to find the Maximum value that covers the dGamma/dE function,
// so simply we scan the function to find the maximum
const int N_E = 200;

// Kinematical bounds to the final-state electron energy
double E_range_min = mElectron;
double E_range_max = mMuon/2.0;
double E_increment  = (E_range_max-E_range_min) / N_E;

// The maximal value to bound the function
double r_max = 0;
for (Int_t ie = 0; ie < N_E; ie++){
  double E_ = E_range_min + ie * E_increment;
  double r_ = dGamma_dE(E_);
  if (r_ > r_max)
      r_max = r_;
} // Done with E scan

// We apply the acceptance-rejection method criteria
double t = r_max * ran->Rndm();
bool  accept = (t < r);
if(accept){
```

```
// Fill the accepted electron energies into the histogram
hE_Electron->Fill(EE1*1000, weight); // GeV->MeV
}
```

Of course, we also need to evaluate the differential decay width for each randomly generated final-state electron energy, $d\Gamma/dE_e(E)$ (see Eq. 4.4). In the code snippet 4.4 this is implemented as the function dGamma_dE(E). You should define this function outside the main body of the code, so that it can be called during the rejection sampling procedure.

Code listing 4.4 Implementation of the differential decay width of the muon decay

```
// Differential width
// - Ee electron energy in GeV
double dGamma_dE(double Ee){

 double r = TMath::Power(8*GF/TMath::Sqrt(2),
↪   2.0)*(mMuon*mMuon*Ee*Ee/(2*TMath::Power(4*TMath::Pi(),
↪   3.0)))*(1.0-(4*Ee/(3.0*mMuon)));

  return r;
}
```

The result of executing the event generator with the additional lines from code listing 4.3 is shown as the black histogram in Fig. 4.2. By accepting generated event kinematics through the rejection sampling method using the function $d\Gamma/dE_e$, we not only sample the kinematically allowed phase-space configurations, but also incorporate the correct physical contribution for each event.

4.1.1   Weighted and Unweighted Events

The reader may have noticed that we apply event weights—specifically, phase-space factors—when plotting distributions. For example, to fill a histogram with the final-state electron energy in a decay, we use

```
hE_Electron->Fill(E, weight);
```

Here, the second argument to the Fill function is the phase-space weight. Rather than simply incrementing the histogram bin corresponding to energy E by one, this line increases the bin content by the value of weight, effectively giving more statistical weight to phase-space configurations that occur more frequently.

What is the rationale behind applying these weights? A short answer is: efficiency. When we first generated muon decay events—without using rejection sampling, as shown in code listing 4.1—we accepted *all* events proposed by TGenPhaseSpace. These events, however, are not uniformly distributed in physical phase space; they follow a distribution determined by the volume element of the three-body final state. The weights applied in the histogram are essentially proportional to these phase-space factors, which also appear under the integral sign in the calculation of the total decay width.

These weights reflect the physical probability density of the corresponding kinematic configuration. But they can also be viewed from a practical perspective: Using event weights allows us to bypass the need for an explicit (and sometimes inefficient) rejection sampling algorithm. Even if a sampled configuration is unlikely, we still keep it—but assign it a small weight. This way, all generated configurations contribute to the final distribution, and the statistical sampling becomes more efficient, especially in high-dimensional phase spaces.

However, this gain in efficiency comes at a cost: *Every* subsequent calculation or histogram that uses the generated events must incorporate these weights. In code listing 4.3, for example, we used the weights in both plotting and computing observables. This additional step can be cumbersome, and for this reason, many event generators prefer to use so-called *unweighted events*. Another subtlety is that some event generators may not sample the event kinematics uniformly. In such cases, one must take care when normalizing events or computing integrals—simple formulas valid only for uniform sampling over the domain may not yield accurate results.

Unweighted events can be generated from weighted events by applying a rejection sampling algorithm. Instead of using a differential decay width or other physical functions to define the acceptance criterion, the acceptance probability is set by the ratio of each event's weight to the maximum event weight observed in the sample. In this way, we retain only those events that are statistically representative of the underlying physical distribution, and each accepted event is counted with equal weight in subsequent analyses. Specifically, this means using the maximum event weight from TGenPhaseSpace via double wmax = event.GetWtMax() as the value for a constant bounding function of the target distribution, which itself is the distribution provided by the event weights via double weight = event.GetGenerate(). Then the same logic of rejection sampling can be used by sampling a uniform random number and accepting the event with an appropriate probability, formally:

```
...
double wmax = event.GetWtMax();
double safety_factor = 1.2;
wmax *= safety_factor;

bool accept_ev = false;
while(!accept_ev){
        double weight = event.Generate();
        accept_ev = (wmax * ran->Rndm() <=weight);
        if(accept_ev){
```

```
            // Implement another rejection sampling based on the further physical
            // functions, dGamma/dE, dSigma/dE, etc.
      ...
            }
      ...
      }
```

4.1.2 Writing an Event Generator from Scratch

To gain further insight, we now consider how to write an event generator for muon decay *without* relying on `TGenPhaseSpace`. Instead, we follow the classical phase-space generation procedures outlined by James [15] and Block [16].

Let us first consider the two-body decay process, $P \rightarrow p_1 + p_2$, in the rest frame of the parent particle, where $P = (M, 0, 0, 0)$. We already established that, in this frame, the two final-state particles must be emitted back to back to conserve momentum, and the invariant mass of the parent must equal the total energy of the final state. From basic kinematics, we find

$$(P - p_1)^2 = M^2 + m_1^2 - 2M E_1 = m_2^2,$$

$$E_1 = \frac{M^2 + m_1^2 - m_2^2}{2M},$$

$$|\mathbf{p}_1| = \sqrt{E_1^2 - m_1^2}.$$

We then generate a random direction (θ_1, ϕ_1) for the three-momentum vector $\mathbf{p}_1$ using

$$\cos \theta_1 = 2 r_\theta - 1,$$

$$\phi_1 = 2\pi r_\phi,$$

where r_θ and r_ϕ are uniformly distributed random numbers in the interval $[0, 1]$. This defines the four-momentum:

$$p_1 = (E_1, \, |\mathbf{p}_1| \sin \theta_1 \cos \phi_1, \, |\mathbf{p}_1| \sin \theta_1 \sin \phi_1, \, |\mathbf{p}_1| \cos \theta_1).$$

The second particle's four-momentum is then fixed by conservation of four-momentum: $p_2 = P - p_1$.

As discussed previously, a three-body decay $P \rightarrow p_1 + p_2 + p_3$ can be constructed by modelling it as a sequence of two-body decays:

$$P \rightarrow P_{12} + p_3, \quad P_{12} \rightarrow p_1 + p_2,$$

where $M_{12} = \sqrt{(p_1 + p_2)^2}$ is the invariant mass of the intermediate particle P_{12}. The kinematically allowed range for M_{12} is constrained by

$$m_1 + m_2 < M_{12} < M - m_3. \tag{4.5}$$

We can sample a value of M_{12} by generating a random number $r \in [0, 1]$ and assigning:

$$M_{12} = r(M - m_3) + (m_1 + m_2).$$

Next, we perform a two-body decay of $P_{12} \to p_1 + p_2$ in the rest frame of P_{12}, using the method described above. The resulting momenta p_1 and p_2 must then be Lorentz-boosted into the lab frame, where P_{12} itself was moving in the decay $P \to P_{12} + p_3$ (with P at rest). This yields the final set of four-momenta for all three decay products.

We can then compute the phase-space event weights using the expressions discussed earlier and apply rejection sampling using a suitable function (e.g. the leading-order differential decay width). The final result is a correctly weighted phase-space distribution for a three-body decay. See the code listing 4.5 below for a concrete implementation. Note that the rejection sampling part is omitted here, as it follows the same structure as in the previous example.

Code listing 4.5 Example implementation of a simple three-body decay event generator from scratch for the muon decay

```cpp
#include <TLorentzVector.h>
#include <TVector3.h>
#include <TMath.h>

#include "Constants.h"

void EventGenFromScratch(){

    // ROOT random number generator
    TRandom2 * ran = new TRandom2(0);

    // Set the masses for calculation of the PS
    double M = mMuon; // parent
    double m1 = mElectron;
    double m2 = mNu;
    double m3 = mNu;

    // Maximum event weight for 3-body decay
    double p1max = TMath::Sqrt((M*M - (m1+m2+m3)*(m1+m2+m3))*(M*M -
    ↪   (m1+m2-m3)*(m1+m2-m3)))/(2*M);
    double p2max = TMath::Sqrt(((M-m3)*(M-m3) -
    ↪   (m1+m2)*(m1+m2))*((M-m3)*(M-m3) - (m1+m2)*(m1+m2)))/(2*(M-m3));
    double weightmax = p1max*p2max;
```

```cpp
// Histogram for the electron energy distribution
TH1F * hE = new TH1F("hE", "electron E", 250, 0, 100);

// Decaying parent particle at rest
TLorentzVector P(0, 0, 0, M);

// Main event loop
Int_t Nevents = 1e+05;
for(int i = 0; i < Nevents; i++){

  // Generate decay: M -> M12 and m3
  // Randomly generate M12
  double M12 = ran->Rndm()*(M-m3) + (m1+m2);

  // Assign angles
  double costheta1 = 2*ran->Rndm()-1;
  double theta1 = TMath::ACos(costheta1);
  double phi1 = 2*TMath::Pi()*ran->Rndm();

  // Get P12 and P3
  double E12 = (M*M + M12*M12 - m3*m3)/(2*M);
  double p12 = TMath::Sqrt(E12*E12 - M12*M12);
  TLorentzVector P12(p12*TMath::Sin(theta1)*TMath::Cos(phi1),
↪ p12*TMath::Sin(theta1)*TMath::Sin(phi1), p12*costheta1, E12);
  TLorentzVector P3 = P - P12; // neutrino

  // Boost into the rest-frame of P12
  TVector3 beta_lab_to_rest = P12.BoostVector();
  TLorentzVector P12_rest = TLorentzVector(P12);
  P12_rest.Boost(-beta_lab_to_rest);

  // Decay P12 into m1 and m2
  // Assign angles
  double costheta = 2*ran->Rndm()-1;
  double theta = TMath::ACos(costheta);
  double phi = 2*TMath::Pi()*ran->Rndm();

  // Get P1 and P2
  double E1 = (M12*M12 + m1*m1 - m2*m2)/(2*M12);
  double pmom1 = TMath::Sqrt(E1*E1 - m1*m1);
  TLorentzVector P1(pmom1*TMath::Sin(theta)*TMath::Cos(phi),
↪ pmom1*TMath::Sin(theta)*TMath::Sin(phi), pmom1*costheta, E1); //
↪ electron
```

```
TLorentzVector P2 = P12_rest - P1;// neutrino

// Boost back P1 and P2 into the rest frame of the decaying particle
P1.Boost(beta_lab_to_rest);
P2.Boost(beta_lab_to_rest);

// Calculate the event weight
double p1 = TMath::Sqrt((M*M - (M12+m3)*(M12+m3))*(M*M -
↪ (M12-m3)*(M12-m3)))/(2*M);
double p2 = TMath::Sqrt(((M12*M12) - (m1+m2)*(m1+m2))*((M12*M12) -
↪ (m1+m2)*(m1+m2)))/(2*M12);
double weight = (p1*p2/weightmax);

// Fill the histogram in MeV
hE->Fill(P1.E()*1000, weight);

}

// Draw the distribution
hE->Draw("hist");
hE->GetXaxis()->SetTitle("E [MeV]");

}
```

4.2 Free Neutron Beta Decay

In this section, we study the decay of a free neutron, which is also a three-body decay: $n \rightarrow p^+ + e^- + \bar{\nu}_e$, see Fig. 4.4. The investigation of neutron decay— particularly within atomic nuclei—has deep historical roots, closely tied to the discovery of the neutrino. It was the observation of continuous beta spectra in nuclear decays that first motivated the proposal of a new, light, neutral particle.

Fig. 4.4 Quark-level Feynman diagram of neutron beta decay: $d \rightarrow u + W^-$, then $W^- \rightarrow e^- + \bar{\nu}_e$. The other quarks act as spectators

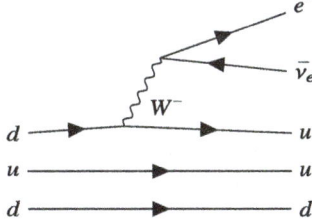

Many radioactive nuclei were already known to emit monoenergetic electrons, corresponding to well-defined energy transitions between nuclear states. However, in certain beta decays, experiments revealed a continuous energy distribution for the emitted electrons. This unexpected observation posed a serious problem for the conservation of energy and momentum in two-body decays.

To see why this was puzzling, consider the two-body decay hypothesis: $n \rightarrow p^+ + e^-$. Assume the neutron is initially at rest, with four-momentum $p_n = (m_n, 0, 0, 0)$. The proton and electron must then recoil in opposite directions, with four-momenta $p_p = (E_p, 0, 0, p_z)$ and $p_e = (E_e, 0, 0, -p_z)$, respectively. Applying four-momentum conservation:

$$p_n = p_p + p_e,$$

$$p_p = p_n - p_e,$$

$$m_p^2 \simeq m_n^2 - 2p_n \cdot p_e = m_n^2 - 2m_n E_e,$$

$$E_e = \frac{m_n^2 - m_p^2}{2m_n} \simeq m_n - m_p \simeq 1.29 \, \text{MeV}. \tag{4.6}$$

This result implies that the final-state electron would always have a fixed energy of about 1.29 MeV—i.e. a monoenergetic line—if the decay were truly two body. However, experiments clearly showed a continuous energy spectrum,[2] leading W. Pauli to propose the existence of an additional, undetected particle: the neutrino.

As we will demonstrate using a phase-space event generator, the inclusion of this third, invisible final-state particle naturally leads to a continuous distribution for the electron energy. This example underscores the crucial role of phase-space considerations: Even when the underlying dynamics of a process are not fully known, phase-space generators can still yield meaningful insight into the expected distributions of observables.

To obtain the physically correct kinematics for neutron decay, however, the known weak interaction dynamics must be incorporated. Since the process closely resembles muon decay, we do not repeat the full derivation of the three-body phase-space formula or its dimensional analysis here. Instead, following Griffiths [1], we use the analytic expression for the differential decay width with respect to the electron energy (with all other final-state variables integrated out), given by

$$\frac{d\Gamma}{dE} = \frac{1}{\hbar c^2 (4\pi)^3} \left(\frac{8G_F}{\sqrt{2}} \right)^2 \frac{1}{4} (c_V^2 + 3c_A^2) J(E) \tag{4.7}$$

with

---

[2] Observed by English physicist J. Chadwick (1891–1974).

$$J(E) = \frac{1}{2}(m_n^2 - m_p^2 - m_e^2)c^2(E_+^2 - E_-^2) - \frac{2m_n}{3}(E_+^3 - E_-^3)$$

$$E_\pm = \frac{\frac{1}{2}(m_n^2 - m_p^2 + m_e^2)c^2 - m_n E}{m_n - E/c^2 \mp |\mathbf{p}|/c}$$

where again G_F is the Fermi coupling constant, and m_n, m_p, and m_e are the masses of the neutron, proton, and electron, respectively. The symbols E and $\mathbf{p}$ refer to the final-state electron's energy and momentum. The constants c_V and c_A are the weak vector and axial-vector coupling constants, respectively, which parametrize the structure of the hadronic vertex $n \rightarrow p + W$ in the weak interaction.

It is straightforward to implement this differential decay width function in code. Since it depends only on the electron energy (with all other final-state degrees of freedom integrated out), the total decay width—and thus the neutron lifetime—can be computed numerically via a one-dimensional integral. This can be efficiently evaluated using standard numerical integration libraries, such as GSL.

See the following code listing 4.6 for a concrete example of how this can be implemented.

Code listing 4.6 Code listing for the GSL integration of the neutron decay differential width, $d\Gamma/dE$

```cpp
#include <TMath.h>
#include "Math/Functor.h"
#include "Math/Integrator.h"
#include "Constants.h"

#include <iostream>


double J(double Ee){

  double Eplus = (0.5*(mNeutron*mNeutron - mProton*mProton +
↪    mElectron*mElectron)
                  - mNeutron*Ee)/(mNeutron - Ee - TMath::Sqrt(Ee*Ee -
↪    mElectron*mElectron));
  double Eminus = (0.5*(mNeutron*mNeutron - mProton*mProton +
↪    mElectron*mElectron)
                  - mNeutron*Ee)/(mNeutron - Ee + TMath::Sqrt(Ee*Ee -
↪    mElectron*mElectron));
  return 0.5*(mNeutron*mNeutron - mProton*mProton -
↪    mElectron*mElectron)*(Eplus*Eplus - Eminus*Eminus)
    - 2*mNeutron*(Eplus*Eplus*Eplus - Eminus*Eminus*Eminus)/3.0;
}
```

```cpp
// Differential width
// - Ee electron energy in GeV
double dGamma_dE(double Ee){
  double r = TMath::Power(8*GF/TMath::Sqrt(2),
↪ 2.0)*(1.0/(TMath::Power(4*kPi, 3.0)))*J(Ee) * (0.25*(cV*cV +
↪ 3*cA*cA));
  return r;
}


void NeutronLifeTime(){

  // Set default tolerances for all integrators
  ROOT::Math::IntegratorOneDimOptions::SetDefaultAbsTolerance(1.E-6);
  ROOT::Math::IntegratorOneDimOptions::SetDefaultRelTolerance(1.E-6);

  // Configure the integration
  ROOT::Math::Functor1D wf(&dGamma_dE);
  ROOT::Math::Integrator
↪ ig(ROOT::Math::IntegrationOneDim::kADAPTIVESINGULAR);
  ig.SetFunction(wf);
  double E_range_min = mElectron;
  double E_range_max = mNeutron - mProton;
  double val = ig.Integral(E_range_min,E_range_max);
  std::cout << "Gamma from GSL integral: " << val << " GeV, lifetime: "
↪ << (1/val)*perGeV_to_s << " sec" << std::endl;

}
```

Code listing 4.7 Example console output of the code listing 4.6

```
root [0] .x NeutronLifeTime.C
Gamma from GSL integral: 7.19256e-28 GeV, lifetime: 896.923 sec
```

We obtain a numerical value for the neutron mean lifetime of approximately 897 seconds. This is somewhat larger than the experimentally measured mean value of 877.75 ± 0.28 seconds reported by one of the most recent high-precision measurements [2].

This discrepancy is expected. One of the key challenges in accurately modelling neutron beta decay lies in the semileptonic nature of the process: The W boson couples to a hadronic system at the neutron-proton vertex. In contrast, muon decay is a purely leptonic process, where the weak interaction is well described by the $(1 - \gamma^5)$ V-A structure in the theory. For neutron decay, the hadronic current is only approximately captured by the inclusion of effective vector and axial-vector couplings, c_V and c_A, respectively. These introduce important—but simplified— corrections to the weak vertex in the context of this leading-order calculation.

As in the case of the muon three-body decay, we can write a Monte Carlo code to simulate the final-state kinematics of free neutron decay using a phase-space generator and apply rejection sampling. We compare the electron energy spectrum obtained from the pure phase-space distribution to the one derived using the physically motivated differential decay width via rejection sampling.

In the code listing 4.8 below, we show only the main function. The auxiliary functions used to evaluate the differential decay width are identical to those introduced in the previous muon decay example.

Code listing 4.8   Code listing for generating the final-state phase space for the electron from the free neutron decay at rest

```
// ....
#include <TH1F.h>
#include <TMath.h>
#include <TGenPhaseSpace.h>
#include <TRandom2.h>
#include <TLorentzVector.h>
#include "Constants.h"

#include <iostream>

void NeutronDecay(){
  // Random number generator
  TRandom2 * ran = new TRandom2();

  // Prepare histograms
  TH1F *hE_Electron = new TH1F("hE_Electron","Electron E", 200,0,1.75);
  TH1F *hE_ElectronPS = new TH1F("hE_ElectronPS","Electron E from PS",
↪   200,0,1.75);

  // Main event loop
  unsigned int Nev = 1e+6;
  for (unsigned int nEv=0;nEv<Nev;nEv++) {

    // Neutron at rest  -> proton + electron + nu
    TLorentzVector pNeutron(0.0, 0.0, 0.0, mNeutron);
    Double_t masses[3] = { mProton, mElectron , mNu};

    // Generate the Decay in then nucleon at rest frame
    TGenPhaseSpace event;
    bool allowed = event.SetDecay(pNeutron, 3, masses);
```

```
if(!allowed){
  cout << "Not allowed!" << endl;
  continue;
}
Double_t weight = event.Generate();

// Get the decay daughter particles
TLorentzVector *pProton = event.GetDecay(0);
TLorentzVector *pElectron = event.GetDecay(1);
TLorentzVector *pNu = event.GetDecay(2);
// Get the electron energy
double EEl = pElectron->E();
// Get the neutrino energy
double Enu = pNu->E();
// Fill the phase-space only histogram
hE_ElectronPS->Fill(EEl*1e+03, weight); // convert GeV to MeV


// Apply the Rejection Sampling algorithm

// Evaluate the differential decay-width function
// at this electron energy
double r = dGamma_dE(EEl);

// We need to find the Maximum that bounds this function dGamma_dE()
const int N_E = 200;
double E_range_min = mElectron;
double E_range_max = mNeutron - mProton;
double E_at_r_max = 0.;
double E_increment = (E_range_max-E_range_min) / N_E;
double SafetyFactor = 1.5; // to cover dGamma/dE fully in the Rej.
Sam.

double r_max = 0;
for (int ie = 0; ie < N_E; ie++){
  double E_ = E_range_min + ie * E_increment;
  double r_ = dGamma_dE(E_);
  if (r_ > r_max){
    E_at_r_max = E_;
    r_max = r_;
  }
} // Done with E scan to find maximum

r_max *= SafetyFactor;
```

```
// Apply the Acceptance-Rejection criterion
bool  accept = (ran->Rndm() < r/r_max);

if(accept){
  // Fill the phase-space + dynamics histogram
  hE_Electron->Fill(EEl*1e+03, weight); // convert GeV to MeV
}
}// End of the Event loop

// Plot the results
hE_Electron->Draw("hist");
hE_Electron->GetXaxis()->SetTitle("Electron E [MeV]");
hE_Electron->GetYaxis()->SetTitle("Events");
hE_ElectronPS->Scale(hE_Electron->Integral()/hE_ElectronPS->
Integral());
hE_ElectronPS->Draw("histsame");
hE_ElectronPS->SetLineColor(kRed);
}
```

In code listing 4.8, we begin by creating a random number generator object pointer and preparing histograms to record the final-state electron energy distribution. Within a standard event loop, we construct the four-momentum of a neutron at rest, define the masses of the final-state particles, and follow steps similar to the muon decay case to generate phase-space configurations. We store the phase-space weights and obtain pointers to the final-state four-momentum objects.

Next, we apply the rejection sampling algorithm to accept or reject events based on the physical differential decay width function. To implement this, we must first determine an appropriate upper bound on the decay width. We do this via an internal loop that scans the generated events to estimate the maximum value of the decay width. A safety factor is then applied to this estimate to ensure a conservative upper bound. Once this is set, we perform the actual acceptance test: If the event passes, the corresponding electron energy value is used to fill the appropriate histogram, scaled by the event's phase-space weight.

The output of the code shows two distributions, see Fig. 4.5: one based solely on phase-space weights, and another using phase-space weights combined with rejection sampling. The two resulting histograms exhibit clear differences. In particular, the shape near the upper endpoint of the energy spectrum, $E \simeq 1.29$ MeV, is highly sensitive to the neutrino mass. This endpoint corresponds to the physical limit where nearly all the decay energy is transferred to the electron, leaving the neutrino at rest. In the present simulation, the neutrino is assumed to be massless. However, including a nonzero neutrino mass would slightly distort the spectral shape near the endpoint.

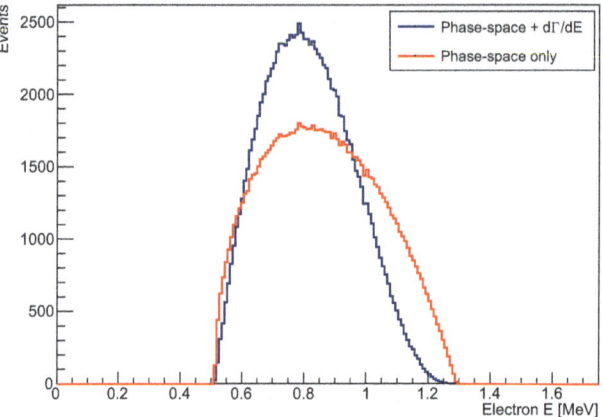

Fig. 4.5 Electron energy distribution from a free neutron decay at rest, showing the distribution with phase-space weights only (red) and with the additional use of the rejection sampling algorithm (blue) to incorporate the differential decay width

In experimental settings, free neutrons are challenging to use for such precise measurements. Instead, experiments often rely on the beta decay of tritium, which allows a clean determination of the spectral shape and endpoint energy, and thus provides sensitivity to the neutrino mass.

4.2.1 Tritium Lifetime

For a rough estimation of the lifetime of light nuclei, we can reuse the previously developed formalism with only slight modifications. Specifically, we adjust the masses of the neutron and proton to reflect their bound-state values within nuclei, and we include the Fermi function,[3] $F(Z', E)$, as a multiplicative correction factor. This function accounts for the Coulomb interaction between the final-state nucleus and the outgoing charged lepton, typically contributing a small correction (close to 1) to the decay rate. A more detailed discussion of the Fermi function is presented in the next chapter.

As an illustrative example, we consider the beta decay of tritium ($^3_1$H), a rare radioactive isotope of hydrogen. The tritium nucleus, known as the triton, consists of one proton and two neutrons. One of the neutrons undergoes beta decay, but as it is bound within the nucleus, we must account for the nuclear binding energy in our calculation.

To do this approximately, we introduce a *mass defect* correction. The effective mass of the decaying neutron is reduced by its binding energy in tritium, $E_b(^3_1H) \simeq$

---

[3] Named after Italian-American physicist E. Fermi (1901–1954).

8.482 MeV [5]. For the final-state proton, which remains bound within the daughter nucleus helium-3 ($^3_2$He), we similarly adjust its mass using the proton's binding energy in $^3_2$He, $E_b(^3_2$He$) \simeq 7.7186$ MeV.

In addition, we multiply the differential decay width by the Fermi function $F(Z', E)$ to account for Coulomb effects. With these modifications in place, we can numerically integrate the same differential decay width formula as used for the free neutron decay. This results in an approximate tritium half-life of $\tau_{1/2} \simeq 11$ years, slightly shorter than the experimentally measured value of about 12.3 years. However, it must be emphasized that this is only a crude approximation and does not include nuclear structure corrections or more sophisticated modelling of the weak transition matrix element.

Tritium is widely used in experiments designed to determine the absolute mass of the neutrino. This is due to the low maximum energy (endpoint) of the final-state electron in tritium beta decay, which is around 18.6 keV. The reader is encouraged to implement this calculation and generate the final-state electron energy spectrum using the methods described earlier, modified for the bound-state masses and including the Fermi function.

Recall that the endpoint of the electron energy spectrum corresponds to the configuration in which the neutrino is produced at rest. Because the available energy is so low near the endpoint, the shape of the spectrum becomes particularly sensitive to the mass of the neutrino—making tritium decay an ideal probe for such measurements.

4.2.2 Nuclear Beta Decay

While the previously discussed neutron decay considered a free neutron, in reality, neutrons are typically bound within atomic nuclei. As a result, nuclear corrections must be applied to the decay spectrum. Using Fermi's golden rule, the transition rate for nuclear beta decay can be written as

$$\Gamma_{fi} = 2\pi |V_{fi}|^2 \rho(E_f), \qquad (4.8)$$

where V_{fi} is the transition matrix element, which includes the initial and final *nuclear* states, and $\rho(E_f)$ is the density of final states. The matrix element V_{fi} involves the spatial integration over the interaction potential and the wave functions of the participating particles—namely the nucleus, the electron, and the neutrino. Formally, it is given by

$$V_{fi} = G_\beta \sum_i \int \left(\Psi_f^* \phi_e^* \phi_\nu^* O_i \Psi_i \right) d^3\mathbf{x}. \qquad (4.9)$$

Here, $\Psi_{f,i}$ denote the nuclear final and initial-state wave functions, $\phi_{e,\nu}$ are the electron and neutrino wave functions, $G_\beta = G_F|V_{ud}|$ is the effective weak interaction strength in beta decay, with V_{ud} the u-d quark mixing matrix element, and O_i represent the various operators characterizing the nuclear interaction (the details of which lie beyond the scope of this book). The matrix element is commonly factorized as $V_{fi} \propto G_\beta M_{fi}$, where M_{fi} is the so-called nuclear matrix element.

After some algebraic manipulation (see, e.g. [14] for a textbook derivation), it can be shown that the shape of the final-state electron energy spectrum for the so-called *allowed transitions*[4] is primarily determined by the density of final states. Consequently, the differential electron energy spectrum takes the form

$$\frac{d\Gamma}{dE_e} \propto p_e E_e p_\nu^2, \tag{4.10}$$

where p_e and p_ν are the momenta of the electron and neutrino, respectively, and E_e is the electron energy. As before, the endpoint of the electron energy spectrum, E_0, corresponds to the case when the neutrino carries its minimum energy, i.e. its rest mass m_ν. Expressing the neutrino energy in terms of the endpoint energy as $E_\nu = (E_0 - E_e) + m_\nu$, one finds the squared neutrino momentum to be

$$p_\nu^2 = E_\nu^2 - m_\nu^2 = (E_0 - E_e)^2 + 2m_\nu^2(E_0 - E_e) = (E_0 - E_e)(E_0 - E_e + 2m_\nu). \tag{4.11}$$

Thus, the electron energy spectrum becomes sensitive to the neutrino mass:

$$\frac{d\Gamma}{dE_e} \propto p_e E_e(E_0 - E_e)(E_0 - E_e + 2m_\nu). \tag{4.12}$$

However, the full β spectrum includes additional correction factors and is generally expressed as

$$\frac{d\Gamma}{dE_e} \propto p_e E_e(E_0 - E_e)^2 F(Z', E_e)|M_{fi}|^2 S(p_e, p_\nu), \tag{4.13}$$

where $F(Z', E_e)$ is the Fermi function [6], which accounts for the Coulomb interaction between the daughter nucleus (of charge Z') and the outgoing electron, and $S(p_e, p_\nu)$ encodes corrections related to the forbiddenness of the transition.

In the regime where $(\alpha Z)^2 \ll 1$, with $\alpha \simeq 1/137$ being the fine-structure constant, the Fermi function can be approximated by

$$F(Z', E_e) = \frac{2\pi y}{1 - e^{-2\pi y}}, \tag{4.14}$$

---

[4] Transitions are classified as allowed if the combined lepton orbital angular momentum is $L = L_e + L_\nu = 0$ or 1. In these cases, the spin angular momenta of the electron and neutrino may be parallel or antiparallel. There also exist *forbidden* transitions, which are not strictly forbidden but are suppressed due to higher angular momentum configurations, resulting in longer lifetimes.

where $y = \alpha Z E_e / p_e$. For the daughter nucleus helium-3, this correction factor is close to 1.2. In general, the Fermi correction approaches unity as the nuclear charge Z decreases.

4.3 Solar Neutrinos

Star formation [3] results from the gravitational collapse of matter—primarily hydrogen and helium—during which part of the energy is released in the form of radiation, while the remaining energy heats the collapsing cloud, maintaining hydrostatic equilibrium. At the high temperatures involved, atoms become fully ionized, and as the cloud contracts further, nuclear processes begin to activate.

One of the key processes is hydrogen burning that transforms protons into ^{4}He nuclei. The net result of the reaction chain is $4p \rightarrow {}^4\text{He} + 2e^+ + 2\nu_e$, which must include weak interaction to transform protons into neutrons, a process similar to inverse beta decay, $p \rightarrow n + e^+ + \nu_e$. Since two protons cannot form a bound state, one proton must convert into a neutron and bind with the other to form a deuteron. This reaction is the initial step in Branch I of the so-called *proton-proton chain*, one of the primary mechanisms of hydrogen burning in the Sun, see Fig. 4.6:

$$p + p \rightarrow d + e^+ + \nu_e \tag{4.15}$$

$$p + d \rightarrow {}^3\text{He} + \gamma \tag{4.16}$$

$${}^3\text{He} + {}^3\text{He} \rightarrow {}^4\text{He} + 2p \tag{4.17}$$

which releases about $Q_{\text{eff}} \approx 26\,\text{MeV}$ for each ^{4}He formed. The outcome of this reaction is the emission of electron neutrinos, which escape the stellar environment with almost no interaction.

This weak process is mediated by a matrix element involving the hadronic wave function overlap between the initial proton pair and the resulting deuteron. At low energies, and to first approximation, the matrix element varies slowly. The reaction cross section is given by [4]

$$\sigma(E) = \frac{S(E)}{E} \exp(-2\pi\eta), \tag{4.18}$$

where E is the kinetic energy of the interacting protons in the centre-of-mass frame, $S(E)$ is the slowly varying astrophysical S-factor, and η is the Coulomb-repulsion parameter,

$$\eta = \frac{Z_1 Z_2 \alpha}{\sqrt{2E/\mu}} = Z_1 Z_2 \alpha \sqrt{\frac{\mu}{2}} \cdot \frac{1}{\sqrt{E}} \approx \frac{0.989\,\text{MeV}^{1/2}}{\sqrt{E} \cdot 2\pi}, \tag{4.19}$$

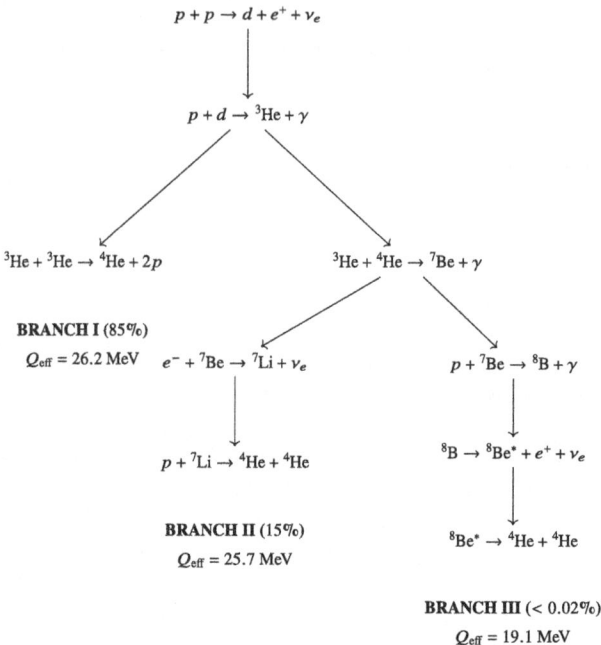

Fig. 4.6 The three branches of the proton-proton chain in the Sun with the effective energy release, Q_{eff}, and termination fraction per branch

with μ the reduced mass of the proton-proton system, and α the fine-structure constant. The exponential factor $\exp(-2\pi\eta)$ is known as the Gamow factor,[5] and it represents the quantum mechanical probability for an s-wave proton to tunnel through the Coulomb barrier and interact with another proton.

At low energies, the S-factor is approximately constant: $S(E) \approx S(0) \approx 4 \times 10^{-22}$ keV·barn [7]. Since $\sigma(E)$ is a slowly varying function of energy, as illustrated in Fig. 4.7, the energy spectrum of the emitted neutrinos is primarily determined by the phase-space distribution.

But how can phase space be generated for a process like $p \rightarrow n + e^+ + \nu_e$, where the energy balance appears reversed compared to a typical decay, given that the proton is lighter than the neutron? The key lies in the binding energy of the deuteron. Although a free proton is approximately 1.8 MeV lighter than a free neutron, the deuteron's binding energy is $E_b(d) \simeq -2.225$ MeV. As a result, the total mass of two free protons is greater than that of a bound proton-neutron pair. This energy difference allows the reaction

$$p + p \rightarrow d + e^+ + \nu_e \tag{4.20}$$

[5] Named after Soviet-American theoretical physicist G. Gamow (1904–1968).

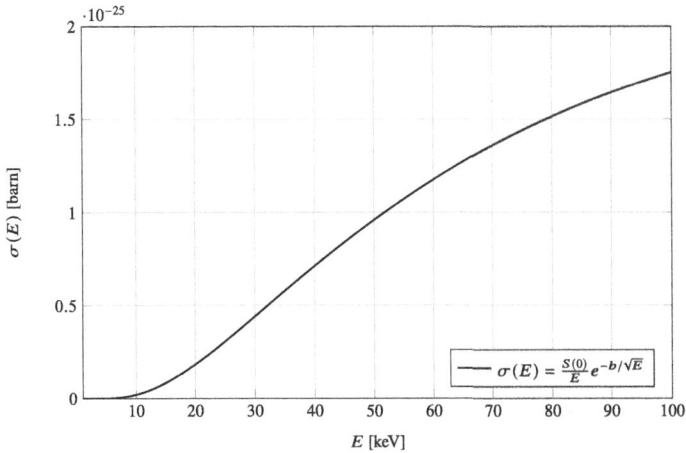

Fig. 4.7 Estimated cross section for proton-proton fusion at low energies, using $S(0) = 4.0 \times 10^{-25}$ MeV·barn and $b = 0.989$ MeV$^{1/2}$ [7]. The temperature at the centre of the Sun is $T \approx 2 \times 10^{7}$ K, with $k_{\mathrm{B}}T = 1.7$ keV, where k_{B} is the Boltzmann constant. To fuse, two protons must penetrate the Coulomb barrier, the probability of which is $P \propto \exp[-(E_G/E)^{1/2}]$, where $E_G = 493$ keV is the Gamow energy. Fusion of protons is most likely at an energy of $E \approx 7$ keV [3]

to proceed, with an excess of energy available in the final state. The maximum energy that can be transferred to the neutrino is approximately 0.42 MeV, which defines the endpoint of the solar neutrino energy spectrum from this process. This allows us to simulate the kinematics by randomly sampling the available phase space, as shown in code listing 4.9.

Code listing 4.9 Generating the final-state phase space for the solar neutrino from $p + p \rightarrow d + e^+ + \nu_e$

```
#include <TH1F.h>
#include <TCanvas.h>
#include <TGenPhaseSpace.h>
#include <TLorentzVector.h>
#include <TRandom.h>

#include "Constants.h"

void PPFusion(){
    // Random number generator
    TRandom * ran = new TRandom2();

    // Prepare histograms
    TH1F *hE_Nu_PP = new TH1F("hE_Nu_PP","Nu E", 2000,0,20.0);
```

```cpp
// Main event loop for pp
unsigned int Nev = 1e+6;
for (unsigned int nEv=0;nEv<Nev;nEv++) {

    // 2*proton  -> deuteron + e^{+} + nu
    TLorentzVector pProton(0.0, 0.0, 0.0, mProton*2);
    Double_t masses[3] = { mDeuteron, mElectron , mNu};

    TGenPhaseSpace event;
    bool allowed = event.SetDecay(pProton, 3, masses);
    if(!allowed){
        cout << "Not allowed!" << endl;
      continue;
    }
    Double_t weight = event.Generate();
    // Get the decay daughter particles
    TLorentzVector *pNeutron = event.GetDecay(0);
    TLorentzVector *pElectron = event.GetDecay(1);
    TLorentzVector *pNu  = event.GetDecay(2);

    // Get the neutrino energy
    double Enu = pNu->E();

    hE_Nu_PP->Fill(Enu*1e+3, weight); // convert GeV to MeV

}

// Plot the results
TCanvas * c = new TCanvas("c", "c", 1);
hE_Nu_PP->Draw("hist");
hE_Nu_PP->SetLineColor(kBlue);
hE_Nu_PP->GetXaxis()->SetRangeUser(0.1, 20);
hE_Nu_PP->GetYaxis()->SetRangeUser(1, 1e+05);
hE_Nu_PP->GetXaxis()->SetTitle("E [MeV]");
// Set Log-Log scale
c->SetLogy();
c->SetLogx();
}
```

The resulting spectrum (alongside others) is shown in Fig. 4.8 as the dashed-dotted histogram. As expected, the endpoint of the neutrino energy distribution appears at 0.42 MeV. Various other processes in the Sun also produce neutrinos—some are two-body decays yielding monoenergetic neutrinos, while others are

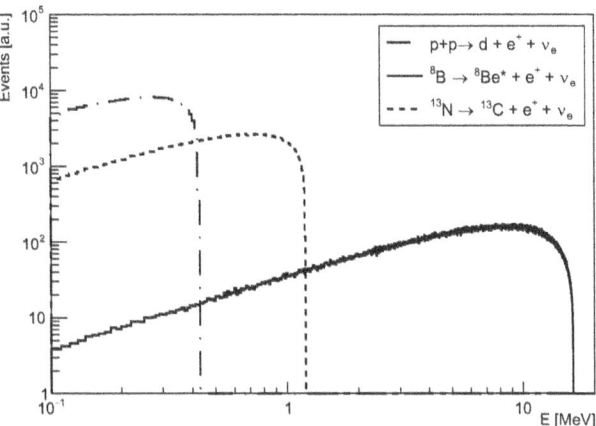

Fig. 4.8 Neutrino energy distribution from the phase-space generator for solar neutrinos

three-body decays that produce continuous spectra. Two additional three-body processes are illustrated in Fig. 4.8, and their phase-space distributions can be generated analogously to the previous example.

One important reaction is the decay

$$^{8}\text{B} \rightarrow {}^{8}\text{Be}^* + e^+ + \nu_e, \tag{4.21}$$

which emits some of the highest-energy neutrinos from the Sun.[6] These neutrinos are especially relevant because they can be detected by terrestrial experiments. To simulate this process, we again use the mass defect method: We subtract the average binding energy of a proton in ^{8}B, $E_b(^{8}\text{B}) \simeq -37.74$ MeV, from its nominal mass, and similarly subtract the binding energy of a neutron in ^{8}Be, $E_b(^{8}\text{Be}) \simeq -56.50$ MeV. This provides a rough estimate of the available energy and allows a naive simulation of the ^{8}B neutrino spectrum using phase-space sampling. This process is part of the extended proton-proton chain.

Another pathway for hydrogen burning involves heavier elements and relies on the presence of carbon, produced in earlier stellar generations. One example is proton capture on carbon:

$$p + {}^{12}\text{C} \rightarrow {}^{13}\text{N} + \gamma, \tag{4.22}$$

which is part of the so-called *CNO cycle*, The net result of the cycle is the synthesis of ^{4}He along with excess energy. As far as neutrino production is concerned, the ^{13}N nucleus can undergo beta decay:

---

[6] The only known process producing even slightly more energetic neutrinos is the so-called *hep* reaction, $^{3}\text{He} + p \rightarrow {}^{4}\text{He} + e^+ + \nu_e$, although neutrinos from this process are extremely rare.

$$^{13}\text{N} \rightarrow {}^{13}\text{C} + e^+ + \nu_e. \tag{4.23}$$

The endpoint energy of the resulting neutrinos (1.2 MeV) can again be estimated using the mass defect method. This involves subtracting the average proton binding energy in ^{13}N, $E_b(^{13}\text{N}) \simeq -94.11$ MeV, from its mass, and similarly accounting for the neutron binding energy in ^{13}C, $E_b(^{13}\text{C}) \simeq -97.11$ MeV. It should be noted that this approximation yields qualitatively reasonable results primarily for nuclei with low atomic number (Z).

4.4 Pion and Kaon Decay, Accelerator Neutrinos

Experimentally, accelerator neutrino beams are produced by directing a high-energy proton beam onto a fixed target, generating a shower of secondary particles, primarily charged mesons (pions and kaons). These mesons are focused using magnetic horns and allowed to decay in a dedicated decay tunnel. Unwanted charged particles are subsequently absorbed in a large beam dump or shielding at the end of the tunnel.

Interestingly, a tree-level calculation of the weak two-body decay of charged pions (or kaons), $\pi^+ \rightarrow \mu^+ + \nu_\mu$, using Feynman rules [1], yields that the kinematic distributions of the decay products are governed mainly by the available phase space. This property makes such decays particularly useful for producing well-characterized neutrino beams. In what follows, we will simulate the phase-space kinematics using TGenPhaseSpace and investigate the energy and angular distributions of neutrinos produced in the in-flight decay of charged pions.

Before turning to the simulation, let us first perform a theoretical calculation of the expected outcome. Consider a charged pion with mass $m_\pi = 139.6$ MeV and momentum aligned along the z-axis, with a magnitude [8]:

$$|\mathbf{p}_\pi| = p_{z,\pi} \equiv k_\pi = 5 \text{ GeV}. \tag{4.24}$$

The charged pion decays into a muon with mass $m_\mu = 105.7$ MeV and a muon neutrino. Let the angle between the pion flight direction and the outgoing neutrino be denoted by ϑ. The relevant kinematics are analysed in the laboratory frame, consider Fig. 4.9.

In the rest frame of the pion, the decay products are emitted back to back. Let us write the components of the four-momenta of the charged pion and the neutrino in a column vector form:

$$p_\pi = \begin{pmatrix} E_\pi \\ k_\pi \\ 0 \\ 0 \end{pmatrix}, \quad p_\nu = \begin{pmatrix} E_\nu \\ k_\nu \cos\vartheta \\ k_\nu \sin\vartheta \\ 0 \end{pmatrix}. \tag{4.25}$$

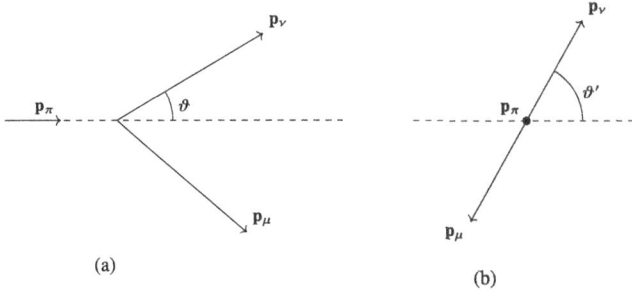

Fig. 4.9 Reference frames for the charged pion decay. (**a**) Laboratory frame. (**b**) Rest frame

Using four-momentum conservation and standard Lorentz-invariant properties, we write

$$p_\pi = p_\mu + p_\nu$$
$$p_\mu^2 = (p_\pi - p_\nu)^2$$
$$m_\mu^2 = m_\pi^2 - 2p_\pi \cdot p_\nu = m_\pi^2 - 2E_\nu(E_\pi - k_\pi \cos \vartheta).$$

From this, we directly obtain the expression for the neutrino energy in the charged pion decay:

$$E_\nu = \frac{m_\pi^2 - m_\mu^2}{2(E_\pi - k_\pi \cos \vartheta)} \tag{4.26}$$

and similarly, for charged kaon decay:

$$E_\nu = \frac{m_K^2 - m_\mu^2}{2(E_K - k_K \cos \vartheta)}. \tag{4.27}$$

The energy of the final-state neutrino from charged pion (or kaon) decay depends on the energy and momentum of the parent meson and the angle ϑ between the meson and neutrino directions. Since kaons are roughly three times more massive than pions, they dominate the high-energy tail of the neutrino spectrum.

In this example, we assume a 5 GeV pion, so the only unknown variable is the emission angle ϑ. Using $E_\pi = \sqrt{m_\pi^2 + k_\pi^2}$, we can compute

$$E_\nu^{\max} = E_\nu(\vartheta = 0) = \frac{m_\pi^2 - m_\mu^2}{2(E_\pi - k_\pi)} = 2.13 \, \text{GeV}$$

$$E_\nu^{\min} = E_\nu(\vartheta = \pi) = \frac{m_\pi^2 - m_\mu^2}{2(E_\pi + k_\pi)} = 0.42 \, \text{MeV} \tag{4.28}$$

$$E_\nu(\vartheta = 2.5°) = \frac{m_\pi^2 - m_\mu^2}{2(E_\pi - k_\pi \cos(2.5°))} = 620 \, \text{MeV}.$$

These values demonstrate that the more the neutrino is emitted in the forward direction (i.e. aligned with the pion), the higher its energy. The angle $\vartheta = 2.5°$ is of particular interest because it corresponds to the off-axis angle used in the Tokai-to-Kamioka (T2K) long-baseline experiment, optimized to produce neutrinos with energies around 600 MeV (discussed in a later chapter).

As noted, one can use relativistic kinematics to derive the angular distribution of the final-state neutrinos in the lab frame. To begin, we compute the neutrino's four-momentum in the pion rest frame, p'_ν, by applying a Lorentz boost from the lab frame with $\beta = |\mathbf{p}_\pi|/E_\pi$ and $\gamma = E_\pi/m_\pi$:

$$p'_\nu = \Lambda p_\nu = \begin{pmatrix} \gamma & -\beta\gamma & 0 & 0 \\ -\beta\gamma & \gamma & 0 & 0 \\ 0 & 0 & 1 & 0 \\ 0 & 0 & 0 & 1 \end{pmatrix} \begin{pmatrix} E_\nu \\ k_\nu \cos\vartheta \\ k_\nu \sin\vartheta \\ 0 \end{pmatrix} = \begin{pmatrix} \gamma E_\nu(1 - \beta\cos\vartheta) \\ \gamma E_\nu(\cos\vartheta - \beta) \\ k_\nu \sin\vartheta \\ 0 \end{pmatrix}$$

$$= \begin{pmatrix} E'_\nu \\ k'_\nu \cos\vartheta' \\ k'_\nu \sin\vartheta' \\ 0 \end{pmatrix}. \tag{4.29}$$

We assume that the charged pion travels along the z-axis, corresponding to the first component of the three vectors. From the transformed four-momentum, we have

$$E'_\nu = \gamma E_\nu(1 - \beta\cos\vartheta)$$
$$k'_\nu \cos\vartheta' = \gamma E_\nu(\cos\vartheta - \beta).$$

Dividing the spatial and energy components, we find the transformation rule for the polar angle:

$$\cos\vartheta' = \frac{\cos\vartheta - \beta}{1 - \beta\cos\vartheta} \tag{4.30}$$

$$\cos\vartheta = \frac{\cos\vartheta' + \beta}{1 + \beta\cos\vartheta'}. \tag{4.31}$$

This is the standard relativistic transformation between angles in the lab and rest frames. Next, we consider that the squared amplitude is constant and the decay is isotropic in the pion rest frame. This implies that the differential decay width in solid angle is uniform: $d\Gamma/d\Omega' = K$, a constant. In spherical coordinates:

$$d\Gamma = K \cdot \sin\vartheta' \, d\vartheta' \, d\phi'$$
$$\Rightarrow d\Gamma = K \cdot 2\pi \sin\vartheta' \, d\vartheta'$$
$$\Rightarrow \frac{d\Gamma}{d(\cos\vartheta')} = -K \cdot 2\pi.$$

To find the angular distribution in the lab frame, we use the chain rule:

$$\frac{d\Gamma}{d\vartheta} = \frac{d\Gamma}{d\cos\vartheta} \cdot \frac{d\cos\vartheta}{d\vartheta} = \left(\frac{d\Gamma}{d\cos\vartheta} \right)(-\sin\vartheta)$$

$$= -\sin\vartheta \cdot \frac{d\Gamma}{d\cos\vartheta'} \cdot \frac{d\cos\vartheta'}{d\cos\vartheta}$$

$$= (K \cdot 2\pi \cdot \sin\vartheta)\frac{d\cos\vartheta'}{d\cos\vartheta}.$$

It is left as an exercise for the reader to confirm that $d\cos\vartheta/d\cos\vartheta' = \gamma^2(1 - \beta\cos\vartheta)^2$. Thus, we arrive at the final result for the angular distribution:

$$\frac{d\Gamma}{d\vartheta} = 2\pi K \cdot \frac{\sin\vartheta}{\gamma^2(1 - \beta\cos\vartheta)^2}. \tag{4.32}$$

The shape of this function is plotted in Fig. 4.10 for several values of $\beta = \{0.75, 0.90, 0.99\}$, using $K = 1$. The distribution clearly shows that the higher the momentum of the pion, the more strongly the neutrino is emitted in the forward direction. This explains the extensive efforts devoted to focusing charged mesons with magnetic horns in accelerator-based neutrino beamlines.

Next, we generate the phase space for charged pion decay using TGenPhaseSpace and extract both the energy and angular distributions of the final-state neutrinos. Although the pion is moving in the laboratory frame, the phase space will be sampled in its rest frame. To mimic experimental conditions slightly, we will not assume monoenergetic pions. Instead, their energies will follow a Gaussian probability distribution, as a crude approximation of experimental spectra [8], with small, independent transverse momentum components added. While this setup is not realistic at all, it suffices to illustrate essential features of accelerator neutrino beams.

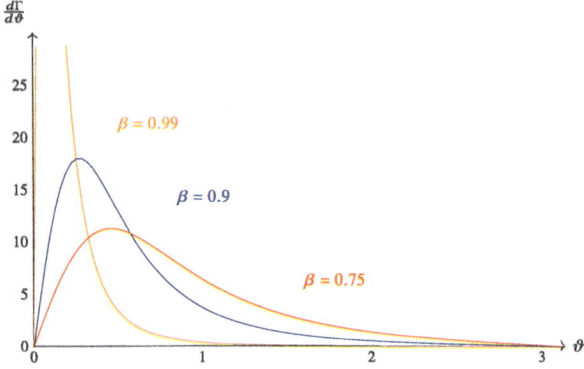

Fig. 4.10 Differential decay width of the charged pion as a function of the neutrino angle

Although ROOT can automate this process, we will implement it manually for clarity, proceeding through the following steps:

- Create a charged pion at rest.
- Assign a laboratory frame energy from a Gaussian distribution.
- Generate its decay products in the pion rest frame.
- Boost the final-state particles into the laboratory frame.
- Plot the neutrino energy and angle distributions, alongside the pion energy distribution.

The code in Listing 4.10 performs these steps. Since the matrix element is constant for this two-body decay, no rejection sampling is needed. Additionally, the code simulates two scenarios within a single run: one fraction of events with $\vartheta = 2.5°$ and the other with $\vartheta = 0°$, corresponding to off-axis and on-axis emission angles between the pion and neutrino.

Code listing 4.10 Code for the phase-space generator for the in-flight charged pion decay with a mean energy of 5 GeV

```
#include <TGenPhaseSpace.h>
#include <TRandom2.h>
#include <TCanvas.h>
#include <TLorentzVector.h>

#include <iostream>

#include "Constants.h"

void MesonDecay(){

  TRandom2 * ran = new TRandom2();

  TH2F * hEnu_Epion = new TH2F("hEnu_Epion", "",100,  0, 10, 150, 0.0,
  ↪  3.0);

  double off_ax = 2.5 * kPi/180.0; // radian
  double off_ax_disc = 0.3 * kPi/180.0; // radian

  double EPion_ = 5.0; // pion energy, GeV

  Int_t Nevents = 1e+07;
  for (Int_t nEv=0;nEv<Nevents;nEv++) {

    // Sample the pion energy from a Gaussian with mean 5, width 2 GeV
    EPion_ = ran->Gaus(5.0, 2.0); // GeV
```

```
if (nEv %3 == 0)
    off_ax = 2.5 * kPi/180.0; // fraction of the events are 2.5 degree
↪ angle
else
    off_ax = 0; // the rest are at 0 deg angle

if(EPion_ < 0) continue; //

// Pion is at rest
TLorentzVector pPion(0.0, 0.0, 0.0, mPion);

// In the lab frame: the pion is moving with some momentum
TLorentzVector pPion_Lab;
pPion_Lab.SetPx(ran->Gaus(0.01, 0.001));// GeV
pPion_Lab.SetPy(ran->Gaus(0.01, 0.001));// GeV
pPion_Lab.SetPz(EPion_); // GeV
pPion_Lab.SetE(TMath::Sqrt(mPion*mPion +
↪ pPion_Lab.Px()*pPion_Lab.Px() + pPion_Lab.Py()*pPion_Lab.Py() +
↪ pPion_Lab.Pz()*pPion_Lab.Pz()));

// Get the boost vector from the Pion four-momentum
TVector3 beta_lab_to_nRest = -(pPion_Lab).BoostVector();
TVector3 beta_nRest_to_lab = -beta_lab_to_nRest;

// Final state particle masses
Double_t masses[2] = { mMuon , mNu };

// Create the decay event in the CM frame
TGenPhaseSpace event;
bool allowed = event.SetDecay(pPion, 2, masses);
if(!allowed)continue;

// Generate the decay
Double_t weight = event.Generate();
TLorentzVector *pMuon = event.GetDecay(0);
TLorentzVector *pNu   = event.GetDecay(1);

// Calculate the angle of the neutrino w.r.t. to the pion direction

// Boost the neutrino and the muon into the Lab frame
TLorentzVector pNu_Lab = TLorentzVector(*pNu);
pNu_Lab.Boost(beta_nRest_to_lab);
TLorentzVector pMuon_Lab = TLorentzVector(*pMuon);
```

```
  pMuon_Lab.Boost(beta_nRest_to_lab);


  // Get the momentum 3-vectors of the Pion and the neutrino in the Lab
↪ frame
  TVector3 beam3V_Lab = pPion_Lab.Vect();
  TVector3 nu3V_Lab = pNu_Lab.Vect();


  // Calculate the difference in the angle between the Pion and the
↪ neutrino
  double theta = nu3V_Lab.Theta();
  double theta_Pion_Lab = pPion_Lab.Theta();
  double theta_nu_Pion = theta - theta_Pion_Lab;


  // Fill the histograms
  double Epion_Lab = pPion_Lab.E();
  double ENu_Lab = pNu_Lab.E();


  // Use the angle between the Pion and neutrino to discriminate events
  // according to its value (a.k.a. off-axis angle)
  if(TMath::Abs(theta_nu_Pion - off_ax) <= off_ax_disc){
    hEnu_Epion->Fill(Epion_Lab, ENu_Lab, weight);
    }
  } // End of event loop

  // Plot the results
  TCanvas * c1 = new TCanvas("c1","c", 1);
  hEnu_Epion->Draw("colsz");
  hEnu_Epion->GetXaxis()->SetTitle("Pion E [GeV]");
  hEnu_Epion->GetYaxis()->SetTitle("Neutrino E [GeV]");
}
```

The script begins by constructing a random number generator, TRandom2.[7] It then defines constants for the particle masses. In the main function, TutorialMesonDecay(), we initialize histograms for various kinematic variables and define the off-axis angle, along with a *discriminator window*, which allows us to classify events based on whether the neutrino emission angle lies near $0°$ or $2.5°$ in the laboratory frame.

To emulate pion production from proton-target collisions, we assign pion energies from a Gaussian distribution centred at 5 GeV. While the actual pion spectrum is not Gaussian [8], this simplification allows us to highlight the energy dependence of neutrinos at different emission angles. We choose a width of $\Delta E = 2\,\text{GeV}$ to

---

[7] If no argument is passed to the constructor, the default linear congruential random number generator uses an initial seed value of 1. See the ROOT documentation for more details.

generate a broad range of pion energies. For event classification, we use the modulus operator: If (`nEv mod 3`) equals zero, the event is assigned $\vartheta = 2.5°$; otherwise, it is treated as on-axis ($\vartheta = 0°$). This provides an efficient way to scan both cases within a single event loop.

The rest of the event loop mirrors earlier examples. We create a four-momentum object for a pion at rest, `pPion`, and for a moving pion in the lab frame, `pPion_Lab`, with $E \approx 5$ GeV. The final pion energy slightly exceeds $5\,\text{GeV}$ due to small transverse momentum fluctuations (on the order of 10 MeV). From the lab-frame four-momentum, we compute the Lorentz boost to transform final-state momenta from the rest frame into the lab frame. Using `Generate()` and `GetDecay()`, we obtain the event weight[8] and final-state kinematics.

To classify events, we compute the polar angle between the outgoing neutrino and the incoming pion in the lab frame. This involves boosting all particles into the lab frame, extracting their three-momenta, and calculating the angle between the pion and neutrino. Events are then separated into *on-axis* ($\vartheta = 0°$) or *off-axis* ($\vartheta = 2.5°$) histograms.

The resulting plots from Listing 4.10 are shown below. The most instructive is Fig. 4.11, which shows the correlation between parent pion energy and daughter neutrino energy at different angles. For on-axis emission ($\vartheta = 0°$), the correlation is nearly linear, allowing a wide energy spread of neutrinos. In contrast, for off-axis emission ($\vartheta = 2.5°$), the neutrino energy spectrum is tightly constrained by kinematics.

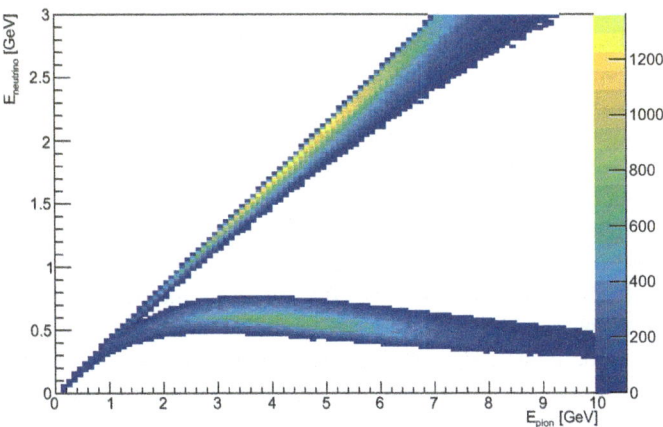

Fig. 4.11 The parent charged pion energy against the daughter neutrino in the laboratory frame, for various angles between them: $\vartheta = 0°$ gives the diagonal band and $\vartheta = 2.5°$ the horizontal band

---

[8] In this case, the weight is constant due to the two-body kinematics and constant matrix element.

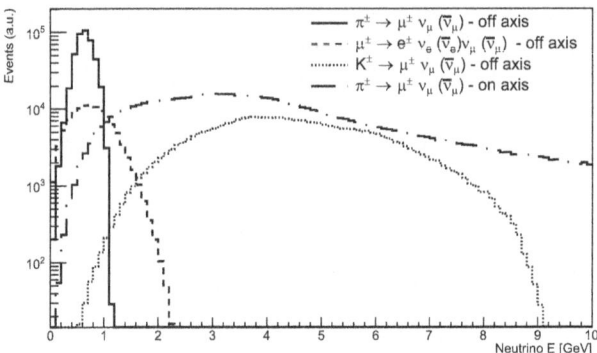

Fig. 4.12 Energy distribution of the final-state neutrino in the laboratory frame at various angles between the parent pion or kaon and the daughter neutrinos, $\vartheta = 0°$ (on axis) and $\vartheta = 2.5°$ (off axis), for the dominant ν_μ and ν_e components

This illustrates a key design feature in accelerator neutrino beams: Experiments such as DUNE [11], which use on-axis configurations, receive a broad neutrino energy spectrum, whereas off-axis experiments such as T2K [12] and the next-generation HyperKamiokande [13] benefit from a narrow-band beam centred at the desired energy. The resulting neutrino energy distributions at on- and off-axis angles are shown in Fig. 4.12 for decays of charged pions, where the pion energy distribution is modelled somewhat more realistically. These results can be directly compared to our earlier analytical estimates of neutrino energies at various angles ϑ, derived in Eq. 4.28 for a fixed 5 GeV pion.

Although our simulation includes a spread in pion energies rather than assuming monoenergetic pions, the observed peak around $E_\nu \simeq 0.6$ GeV in the off-axis ($\vartheta = 2.5°$) distribution aligns well with our previous calculation. Similarly, the on-axis case ($\vartheta = 0°$) shows a broad peak in the neutrino energy spectrum, in qualitative agreement with expectations.

It is important to emphasize that Figs. 4.11 and 4.12 present only a simplified approximation of the actual neutrino energy spectrum. For instance, hadron production details, magnetic horn focusing effects, and geometrical constraints are not included—only phase-space kinematics is modelled. As an exercise, the reader is encouraged to modify the code to replace charged pion decays with those of charged kaons, which results in a significantly harder neutrino spectrum due to the kaon's greater mass.

For a more detailed and realistic modelling of the neutrino beam at T2K, the reader is referred to Refs. [9, 10], which describe the beamline design and hadron production measurements in detail. Finally, Fig. 4.13 shows the distribution of the neutrino-pion angle ϑ in the laboratory frame. The result obtained from our phase-space generator agrees well with the analytical prediction from relativistic kinematics, as shown previously in Fig. 4.10.

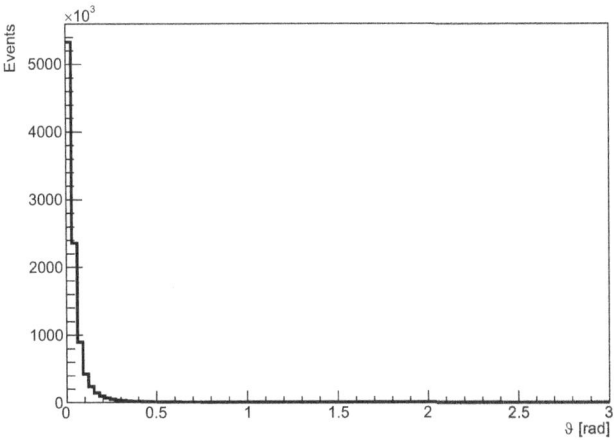

Fig. 4.13 Distribution of the angle, ϑ, see Fig. 4.10 for comparison

References

1. D.J. Griffiths, *Introduction to Elementary Particles*, 2nd edn. (Wiley & Sons, London, 2004)
2. F.M. Gonzalez, Improved neutron lifetime measurement with UCNτ. Phys. Rev. Lett. **127**, 162501 (2021)
3. A.C. Phillips, *The Physics of Starts* (Wiley, London, 1994)
4. J.N. Bahcall, *Neutrino Astrophysics* (Cambridge University Press, Cambridge, 1989)
5. E. Browne, R.B. Firestone, V.S. Shirley, *Table of Radioactive Isotopes* (Wiley, London, 1998)
6. E.J. Konopinski, *The Theory of Beta Radioactivity* (Oxford University Press, Oxford, 1966)
7. E.G. Adelberger et al., Solar fusion cross sections. II. The pp chain and CNO cycles. Rev. Mod. Phys. **83**, 195 (2011)
8. S. Murphy, Measurement of charged pion cross section in proton carbon interaction at 30 GeV with the NA61/SHINE apparatus. AIP Conf. Proc. **1382**, 176–178 (2011)
9. S. Murphy, *Measurement of charged Pion and Kaon production cross sections with NA61/SHINE for T2K*, No. CERN-THESIS-2012-093. CERN-THESIS-2012-429 (2012)
10. A.K. Ichikawa, Design concept of the magnetic horn system for the T2K neutrino beam. NIM A **690**, 27–33 (2012)
11. B. Abi et al., *Deep Underground Neutrino Experiment (DUNE), Far Detector Technical Design Report, Volume II: DUNE Physics* (2020). arXiv:2002.03005
12. The T2K Collaboration, Constraint on the matter–antimatter symmetry-violating phase in neutrino oscillations. Nature **580**, 339–344 (2020)
13. K. Abe et al., Physics potential of a long-baseline neutrino oscillation experiment using a J-PARC neutrino beam and Hyper-Kamiokande. Progr. Theor. Exp. Phys. **2015**, 5 (2015)
14. A. Rubbia, *Phenomenology of Particle Physics* (Cambridge University Press, Cambridge, 2022)
15. F. James, *Monte Carlo Phase Space* (1968) CERN. https://cds.cern.ch/record/275743
16. M.M. Block, Monte Carlo phase space evaluation. Comput. Phys. Commun. **69**, 459–476 (1992)

Neutrino Scattering on Nucleons

5

Abstract

In this chapter, we explore neutrino-nucleon scattering processes that play a central role in accelerator-based neutrino experiments. In particular, we examine two historically significant models used to simulate charged-current quasi-elastic (CCQE) scattering and single-pion resonance production. First, we employ the Llewellyn Smith differential cross section to simulate the phase space of CCQE neutrino-nucleon scattering events, beginning with the simplifying assumption of a free nucleon at rest. We then introduce a basic extension of this model to account for scattering off nucleons in motion inside nuclei. To address inelastic interactions, we illustrate the use of the Rein-Sehgal model in generating charged-current events involving single-pion resonances. This provides a foundation for understanding some key features of neutrino-nucleon interactions relevant to oscillation experiments at intermediate energies.

5.1 Neutrino-Nucleon Scattering Processes at Intermediate Energies

Neutrino-nucleus scattering processes are broadly categorized according to the dominant interaction mechanisms as a function of neutrino energy, see Table 5.1. At the lowest energies (below $\sim$50 MeV), neutrinos interact coherently with entire nuclei through processes such as coherent elastic scattering. In the intermediate energy range ($\sim$100 MeV to a few GeV), interactions with individual nucleons dominate. These include quasi-elastic (QE) scattering, single-pion production via resonance excitation, and coherent pion production. At higher energies (above a few GeV), deep inelastic scattering (DIS) becomes the primary mechanism, where neutrinos probe the partonic substructure of the nucleon.

© The Author(s), under exclusive license to Springer Nature Switzerland AG 2026
B. Radics, *Neutrino Physics*, Lecture Notes in Physics 1043,
https://doi.org/10.1007/978-3-032-03993-4_5

Table 5.1 Typical energy ranges and representative processes for neutrino-nucleus interactions at intermediate energies

Interaction type	Typical energy range	Example process
Coherent elastic scattering (NC)	$\lesssim 50$ MeV	$\nu + A \rightarrow \nu + A$
Charged-current quasi-elastic (CCQE)	~ 0.1–1 GeV	$\nu_l + n \rightarrow l^- + p$
Neutral-current elastic scattering (NC elastic)	~ 0.1–1 GeV	$\nu + N \rightarrow \nu + N$
Single-pion production (resonance, CC/NC)	~ 0.3–2 GeV	$\nu_l + N \rightarrow l^- + N' + \pi$
Coherent pion production (CC/NC)	~ 0.5–2 GeV	$\nu_l + A \rightarrow l^- + A + \pi$
Multi-pion / nonresonant inelastic scattering	$\gtrsim 1$ GeV	$\nu_l + N \rightarrow l^- + N + n\pi$
Deep inelastic scattering (DIS)	$\gtrsim 2$–3 GeV	$\nu_l + N \rightarrow l^- + X$

This chapter focuses on the charged-current quasi-elastic (CCQE) interaction—often referred to as 1p1h (one particle-one hole)—as well as single-pion production via resonance excitation. These processes are particularly relevant in the current generation of accelerator-based long-baseline neutrino oscillation experiments, which operate in the few-GeV energy range.[1]

Neutrinos and antineutrinos can scatter off nucleons via charged-current (CC) or neutral-current (NC) interactions. For example, in CCQE scattering, a neutrino converts a neutron into a charged lepton and a proton: $\nu_l + n \rightarrow l^- + p$. In NC elastic scattering, the neutrino remains unchanged: $\nu + N \rightarrow \nu + N$. The "quasi-elastic" terminology refers to the fact that the target nucleon changes identity (neutron $\leftrightarrow$ proton), but the nucleus may remain largely undisturbed. The other processes—such as resonance production, coherent pion production, and DIS—can also proceed via both CC and NC interactions. Among these, CC interactions are especially useful for flavour identification, since the outgoing charged lepton directly reveals the flavour of the incident neutrino. Moreover, in the ideal CCQE case, the incoming neutrino energy can be reconstructed from the final-state lepton kinematics, assuming the initial nucleon is at rest and no final-state interactions occur.

In realistic experimental environments, however, neutrinos interact with bound nucleons inside nuclei. Nuclear effects—including Fermi motion, binding energy, nucleon correlations, and final-state interactions (FSI) such as pion absorption or rescattering—introduce significant challenges to both energy reconstruction and interaction identification. As the energy increases, multi-pion production, resonance interference, and DIS further complicate the picture.

Modern neutrino event generators such as GENIE [1], NEUT [2], NuWro [3], and GiBUU [4] incorporate detailed models of these effects. While these tools are indispensable for experimental analysis, our goal in this chapter is humbler: to illustrate the basic principles of neutrino-nucleon scattering through simplified sim-

---

[1] Additional nuclear effects, such as two-particle-two-hole (2p2h) excitations, meson exchange currents, and short-range correlations, are also known to contribute significantly at intermediate energies. These are supported by evidence from electron scattering experiments, but fall outside the scope of this book.

ulations. We will examine how to generate kinematic distributions from differential cross sections and how to implement basic models of nucleon motion within nuclei.

Much of our knowledge of nuclear structure and response comes from electron scattering experiments, which have long provided constraints on vector current contributions and nuclear models. Neutrino-nucleus scattering introduces new complexity due to the axial-vector component of the weak interaction and the broad energy distribution of the neutrino flux. Unlike electron beams, which can be tuned to fixed energies, neutrino beams require event-by-event intial-state energy reconstruction, which depends on the final-state topology and the accuracy of nuclear modelling.

In this chapter, we focus on the simulation of CCQE and resonance-induced single-pion production using simplified models. Our treatment is illustrative and avoids the full sophistication required for experimental precision. For a more detailed overview of neutrino interactions with nucleons and nuclei, see [5]. For an even broader review of neutrino cross-section models across a wide energy range, see [6].

5.2 Quasi-elastic Scattering on a Free Nucleon

In this section, we apply the same computational techniques introduced in previous chapters—namely phase-space generation and rejection sampling—but we now turn our attention to scattering rather than decay processes. We first restrict ourselves to the simplest case: two particles in the initial state and two particles in the final state.

A classic and historically important example of such a process is *inverse beta decay*, $\bar{\nu}_e + p \rightarrow e^+ + n$, which was used to detect neutrinos for the first time by Reines and Cowan [7] using a large flux of reactor antineutrinos.[2] Similar two-body scattering processes include the charged-current quasi-elastic (CCQE) interaction, $\nu_l + n \rightarrow l^- + p$, which is central to modern accelerator-based neutrino experiments.

To simulate such interactions, we make a number of simplifying assumptions. First, we assume that the four-momentum of the incoming neutrino is precisely known. This is not true in real experimental conditions, where neutrino beams have broad energy spectra and the incoming energy must be reconstructed from final-state kinematics. Second, we assume that the target nucleon is free and initially at rest. Again, this is a simplification: Nucleons are bound within nuclei and have intrinsic momentum distributions due to Fermi motion. Despite these approximations, this model allows us to build an intuitive understanding of the scattering kinematics and dynamics. As in the decay examples, the final-state distributions are governed by both energy-momentum conservation (kinematics) and the differential cross section (dynamics).

In subsequent sections, we will refine this model by introducing probabilistic treatments of the initial states. These include sampling the neutrino energy from a

---

[2] American physicists C. Cowan (1919–1974) and F. Reines (1918–1998).

Fig. 5.1 Charged-current
quasi-elastic scattering:
$\nu_\mu + n \rightarrow p + \mu^-$, mediated
by a W^+ boson

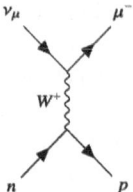

given flux model and the target nucleon's initial momentum from a simple nuclear momentum distribution. These additions will allow us to take a step closer to more realistic simulations of neutrino-nucleon scattering in a nuclear medium.

5.2.1   Constraints from Relativistic Kinematics

Before introducing a numerical model, let us first use relativistic kinematics to examine some physical constraints in the CCQE scattering process. Our working example is

$$\nu_\mu + n \rightarrow p + \mu^-,$$

illustrated in Fig. 5.1, where a neutrino scatters off a neutron at rest, producing a proton and a muon. In experimental analyses, the most important final-state observable is typically the charged lepton—in this case, the muon. We ask two kinematic questions:

1. What is the minimum neutrino energy required for the reaction to occur (the threshold energy)?
2. What are the minimum and maximum values of the final-state muon momentum?

Let us assume the incoming neutrino four-momentum is known and aligned along the z-axis and that the target neutron is at rest. We define the four-momenta in the lab frame as

$$p_\nu = (0, 0, E_\nu, E_\nu), \tag{5.1}$$

$$p_n = (0, 0, 0, M_n), \tag{5.2}$$

$$p_\mu = (|\mathbf{p}_\mu| \sin\theta, 0, |\mathbf{p}_\mu| \cos\theta, E_\mu), \tag{5.3}$$

where θ is the angle between the outgoing muon and the neutrino beam direction, and M_n is the neutron mass.

The threshold condition arises from the requirement that the total centre-of-mass energy $\sqrt{s}$ must be large enough to produce the final-state particles at rest. Using four-momentum conservation,

$$p_\nu + p_n = p_\mu + p_p,$$

we square both sides to obtain:

$$s = (p_\nu + p_n)^2 = M_n^2 + 2p_\nu \cdot p_n = M_n^2 + 2E_\nu M_n, \tag{5.4}$$

since $p_\nu^2 = 0$ and $p_n = (0, 0, 0, M_n)$.

At threshold, the final-state proton and muon are produced at rest in the centre-of-mass frame. Therefore, the minimal value of s corresponds to

$$s_{\mathrm{thr}} = (M_\mu + M_p)^2.$$

Imposing this threshold condition yields

$$M_n^2 + 2E_\nu M_n > (M_\mu + M_p)^2 \tag{5.5}$$

$$\Rightarrow \quad E_\nu > \frac{(M_\mu + M_p)^2 - M_n^2}{2M_n}. \tag{5.6}$$

Substituting numerical values ($M_n = 939.565$ MeV, $M_p = 938.272$ MeV, $M_\mu = 105.658$ MeV), we find $E_\nu^{\mathrm{min}} \simeq 0.109$ GeV.

Another important quantity in neutrino scattering is the momentum transferred to the hadronic final state:

$$q = p_\nu - p_\mu,$$

with its squared value given by $q^2 = -Q^2$, an invariant Lorentz-scalar. The variable Q^2 plays an important role in characterizing different interaction regimes. Importantly, the value of Q^2 is constrained by kinematics, which imposes bounds on the allowed muon energies and momenta. We can express Q^2 in terms of lab-frame quantities by expanding

$$Q^2 = -q^2 = -(p_\nu - p_\mu)^2 = -M_\mu^2 + 2p_\nu \cdot p_\mu$$

$$= -M_\mu^2 + 2E_\nu(E_\mu - |\mathbf{p}_\mu| \cos\theta). \tag{5.7}$$

This result depends on the final-state muon momentum and angle with respect to the incoming beam. Interestingly, the physical limits on Q^2 can be estimated using another quantity, the hadronic invariant mass, W, defined by

$$W^2 = (p_n + q)^2 = (p_n + p_\nu - p_\mu)^2.$$

For quasi-elastic scattering, the only hadron in the final state is the proton, so $W = M_p$. Using four-momentum conservation:

$$W^2 = M_n^2 + 2p_n \cdot p_\nu - 2p_n \cdot p_\mu - Q^2$$
$$= s - 2M_n E_\mu + M_\mu^2 - 2E_\nu(E_\mu - |\mathbf{p}_\mu| \cos\theta), \qquad (5.8)$$

where $s = (p_\nu + p_n)^2 = M_n^2 + 2M_n E_\nu$.

Combining Eqs. 5.7 and 5.8, and expressing everything in terms of invariants, one finds an approximate relationship for Q^2 bounds:

$$Q^2 = -M_\mu^2 + \frac{s - M_n^2}{2s}\left(s + M_\mu^2 - W^2 \pm \sqrt{(s + M_\mu^2 - W^2)^2 - 4sM_\mu^2}\right).$$
$$(5.9)$$

In the limit where the final-state lepton mass is neglected ($M_\mu \to 0$), this simplifies to

$$Q_{max}^2 = \frac{(s - M_n^2)(s - W^2)}{s}, \qquad (5.10)$$

$$Q_{min}^2 = 0. \qquad (5.11)$$

These expressions define the kinematically allowed region for momentum transfer. From Q^2, we can determine the allowed muon energy range. Define the energy transferred to the hadronic system as (see Eq. 5.8)

$$\nu = E_\nu - E_\mu = \frac{p_n \cdot q}{M_n}, \qquad (5.12)$$

$$2M_n\nu = W^2 - M_n^2 + Q^2,$$

from which we extract

$$E_\mu^{min} = E_\nu - \frac{W^2 - M_n^2 + Q_{max}^2}{2M_n}, \qquad (5.13)$$

$$E_\mu^{max} = E_\nu - \frac{W^2 - M_n^2 + Q_{min}^2}{2M_n}. \qquad (5.14)$$

There are also kinematic constraints on W itself. For quasi-elastic scattering, the hadronic invariant mass is simply the proton mass, $W = M_p$. More generally, in resonance or deep inelastic scattering, W can exceed the nucleon mass due to nucleon excitation or multiparticle production. The minimum and maximum values of W are

$$W_{min} = M_{nucleon}, \qquad (5.15)$$

$$W_{max} = \sqrt{s} - M_\mu, \qquad (5.16)$$

where W_{max} corresponds to the scenario in which all remaining energy goes into the hadronic system, leaving the charged lepton at rest. In this limit, neglecting the difference between proton and neutron masses ($M_n \simeq M_p$) and assuming $W = M_p$, we also find from Eq. 5.12:

$$Q^2 \simeq 2M_n \nu. \tag{5.17}$$

Finally, note that all quantities introduced here—s, Q^2, W, and ν—are Lorentz-invariant scalars. This means they can be computed in any frame, though here we have expressed them in the laboratory frame for convenience.

5.2.2 The Llewellyn Smith Charged-Current Quasi-elastic Differential Cross Section

We follow the charged-current quasi-elastic (CCQE) model developed by Llewellyn Smith[3] [8], which assumes scattering on a free nucleon at rest, and we adopt the *impulse approximation*. This approximation treats the incoming neutrino as scattering off a single nucleon within a nucleus of mass number $A = Z + N$, where Z and N are the numbers of protons and neutrons, respectively.

The expression for the differential cross section for a neutrino scattering off a free nucleon at rest is

$$\frac{d\sigma}{dQ^2} = \frac{M^2 G_F^2 \cos^2 \theta_C}{8\pi E_\nu^2} \left[A(Q^2) \mp \frac{(s-u)B(Q^2)}{M^2} + \frac{(s-u)^2 C(Q^2)}{M^4} \right], \tag{5.18}$$

where:

- M is the nucleon mass (typically taken as the neutron mass for $\nu_\mu + n \rightarrow \mu^- + p$).
- G_F is the Fermi constant.
- θ_C is the Cabibbo angle.[4]
- E_ν is the incoming neutrino energy.
- $Q^2 = -q^2$ is the squared four-momentum transfer.
- s and u are Mandelstam variables.[5]
- The $\mp$ sign corresponds to neutrino (upper sign) and antineutrino (lower sign) scattering.

The functions $A(Q^2)$, $B(Q^2)$, and $C(Q^2)$ encode the charge and magnetization distribution of the proton and the neutron, obtained from electron scattering experiments. These include both vector form factors (F_1^V, F_2^V) extracted from

---

[3] Named after British theoretical physicist C. H. Llewellyn Smith (1942-).

[4] Named after Italian theoretical physicist N. Cabibbo (1935–2010).

[5] Named after South African theoretical physicist S. Mandelstam (1928–2016).

electron scattering, the axial form factor(F_A), which is unique to weak interactions, and the pseudoscalar form factor (F_P).

These functions are defined as follows:

$$A(Q^2) = \frac{m_l + Q^2}{M^2} \left(\left[(1+\tau)F_A^2 - (1-\tau)(F_1^V)^2 + \tau(1+\tau))(F_2^V)^2 + 4\tau F_1^V F_2^V \right] - \right.$$

$$\tag{5.19}$$

$$\left. \frac{m_l^2}{4M^2} \left[(F_1^V + F_2^V)^2 + (F_A + 2F_P)^2 - \left(\frac{Q^2}{M^2} + 4 \right) F_P^2 \right] \right)$$

$$B(Q^2) = \frac{Q^2}{M^2} F_A (F_1^V + F_2^V) \tag{5.20}$$

$$C(Q^2) = \frac{1}{4} \left[F_A^2 + (F_1^V)^2 + \tau(F_2^V)^2 \right], \tag{5.21}$$

where $\tau = Q^2/(4M^2)$.

The Mandelstam variables s and u can be written in terms of lab-frame energies. For the process $\nu + n \rightarrow l^- + p$, we have

$$s = (p_\nu + p_n)^2 = M_n^2 + 2E_\nu M_n,$$

$$u = (p_l - p_n)^2 = M_n^2 + m_l^2 - 2E_l M_n,$$

$$s - u = 2M_n(E_\nu + E_l) - m_l^2. \tag{5.22}$$

Alternatively, using the relation $Q^2 \simeq 2M_n(E_\nu - E_l)$ valid for quasi-elastic scattering, we can express this as

$$s - u = 4M_n E_\nu - Q^2 - m_l^2. \tag{5.23}$$

This model forms the foundation for neutrino event generators and cross-section predictions in the CCQE regime.

5.2.3 Form Factors

To parametrize the form factors and define their input values, we follow the approach of [9]. The vector form factors, $F_1^V(Q^2)$ and $F_2^V(Q^2)$, are derived from electron scattering data and are expressed in terms of the Sachs form factors,[6] $G_E^V(Q^2)$ and $G_M^V(Q^2)$. These, in turn, are modelled using the dipole form factor

---

[6] Named after American theoretical physicist R. G. Sachs (1916–1999).

$$G_D(Q^2) = \left(1 + \frac{Q^2}{M_V^2}\right)^{-2} \tag{5.24}$$

with $M_V^2 = 0.71$ GeV^2. Using these definitions, the vector form factors are given by

$$F_1^V(Q^2) = \frac{G_E^V(Q^2) + \frac{Q^2}{4M^2}G_M^V(Q^2)}{1 + \frac{Q^2}{4M^2}} \tag{5.25}$$

$$F_2^V(Q^2) = \frac{G_M^V(Q^2) - G_E^V(Q^2)}{1 + \frac{Q^2}{4M^2}}, \tag{5.26}$$

where the Sachs form factors are obtained from the proton and neutron electric and magnetic form factors:

$$G_E^V(Q^2) = G_E^p(Q^2) - G_E^n(Q^2) \tag{5.27}$$

$$G_M^V(Q^2) = G_M^p(Q^2) - G_M^n(Q^2) \tag{5.28}$$

with individual components given by

$$G_E^p(Q^2) = G_D(Q^2) \tag{5.29}$$

$$G_E^n(Q^2) = 0 \tag{5.30}$$

$$G_M^p(Q^2) = \mu_p\, G_D(Q^2) \tag{5.31}$$

$$G_M^n(Q^2) = \mu_n\, G_D(Q^2). \tag{5.32}$$

Here, μ_p and μ_n are the magnetic moments of the proton and neutron, respectively, in units of the nuclear magneton.[7]

The axial form factor, $F_A(Q^2)$, is also parametrized using a dipole form [10], and the pseudoscalar form factor, $F_P(Q^2)$, is related to the axial one:

$$F_A(Q^2) = g_A\left(1 + \frac{Q^2}{M_A^2}\right)^{-2} \tag{5.33}$$

$$F_P(Q^2) = \frac{2M^2}{Q^2 + m_\pi^2}\, F_A(Q^2), \tag{5.34}$$

---

[7] See the table of constants in Chap. 2.

Table 5.2 Key parameters of the Llewellyn Smith CCQE model used in this book

Symbol	Description	Typical value
G_F	Fermi constant	1.166×10^{-5} GeV^{-2}
$\cos\theta_C$	Cosine of Cabibbo angle	0.974
M	Nucleon mass (target)	0.938 GeV
m_l	Charged lepton mass	For example 0.105 GeV for μ
M_V^2	Dipole mass squared (vector)	0.71 GeV^2
M_A	Axial mass	$\sim$1.0 GeV
g_A	Axial coupling constant	-1.267
m_π	Charged pion mass	0.1396 GeV
μ_p	Proton magnetic moment	2.793
μ_n	Neutron magnetic moment	-1.913

where $g_A = -1.267$ is the axial-vector coupling constant, $M_A^2 \approx 1.0$ GeV^2 is the axial mass,[8] and m_π is the charged pion mass. Table 5.2 summarizes the key parameters of the model used in this book.

5.2.4   Neutral-Current Elastic Scattering

Elastic scattering of neutrinos on nucleons can also occur via the *neutral-current (NC)* weak interaction, mediated by the exchange of a Z^0 boson (see Fig. 5.2). This process is important for understanding neutrino-nucleon interactions, even though it is not used in simulations within this book. The same formalism as in the charged-current case (see Eqs. 5.18–5.19) applies, but the *form factors* are modified to account for the neutral current. In particular, the NC form factors include contributions from strange quark currents. These additional terms appear in both the vector and axial-vector form factors and are typically parametrized with dipole forms as well. The relevant NC vector and axial form factors are

$$F_1^V(Q^2) = \left(\frac{1}{2} - \sin^2\theta_W\right)\left[\frac{\tau_3\left(1 + \tau(1 + \mu_p - \mu_n)\right)}{(1+\tau)\left(1 + Q^2/M_V^2\right)^2}\right]$$

$$- \sin^2\theta_W\left[\frac{1 + \tau(1 + \mu_p + \mu_n)}{(1+\tau)\left(1 + Q^2/M_V^2\right)^2}\right] - \frac{F_1^s(Q^2)}{2} \qquad (5.35)$$

$$F_2^V(Q^2) = \left(\frac{1}{2} - \sin^2\theta_W\right)\left[\frac{\tau_3(\mu_p - \mu_n)}{(1+\tau)\left(1 + Q^2/M_V^2\right)^2}\right]$$

---

[8] The value of M_A can vary slightly between experiments, typically within the range $M_A \approx 0.9$–1.1 GeV.

Fig. 5.2 Examples of neutral-current elastic scattering of neutrinos on nucleons via Z^0 exchange

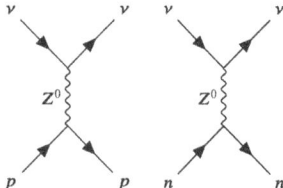

$$- \sin^2 \theta_W \left[\frac{\mu_p + \mu_n}{(1 + \tau) \left(1 + Q^2/M_V^2\right)^2} \right] - \frac{F_2^s(Q^2)}{2} \qquad (5.36)$$

$$F_A(Q^2) = \frac{g_A \tau_3}{2 \left(1 + Q^2/M_A^2\right)^2} - \frac{F_A^s(Q^2)}{2}, \qquad (5.37)$$

where:

- θ_W is the Weinberg angle.
- $\tau = Q^2/(4M^2)$.
- $\tau_3 = +1$ for proton, -1 for neutron.
- μ_p, μ_n are the magnetic moments of the proton and neutron.
- $M_V^2 = 0.71 \, \text{GeV}^2$ is the vector dipole mass squared.
- $g_A = -1.267$, $M_A \approx 1.0 \, \text{GeV}$ is the axial mass.
- $F_1^s(Q^2)$, $F_2^s(Q^2)$, and $F_A^s(Q^2)$ are the strange quark contributions to the vector and axial form factors.

The strange axial form factor is parametrized as

$$F_A^s(Q^2) = \frac{\Delta s}{\left(1 + Q^2/M_A^2\right)^2}, \qquad (5.38)$$

where Δs is the strange quark contribution to the nucleon spin. Since CC interactions do not include F_A^s, the ratio of NC to CC cross sections or that of the differences between neutrino and antineutrino NC and CC cross sections can constrain the strange quark contribution.

5.2.5 Total CCQE Cross Section

Implementing the function in Eq. 5.18 is relatively straightforward for the representative CCQE scattering process, $\nu_\mu + n \rightarrow p + \mu^-$. This function depends on the squared four-momentum transfer, Q^2, and the neutrino energy, E_ν. To evaluate it numerically, some care is required. In particular, the process exhibits a kinematic threshold at $E_{\nu,\text{thr}} \simeq 0.109 \, \text{GeV}$ when the final-state lepton is a muon, as calculated earlier. Below this threshold, the cross section vanishes. A possible implementation of the differential cross section is shown in code listing 5.1. For illustrative purposes, we fix the neutrino energy to $E_\nu = 1$ GeV.

Code listing 5.1   A possible implementation of the Llewellyn Smith charged-current quasi-elastic (CCQE) differential cross section

```cpp
#include <TMath.h>
#include "Math/Functor.h"
#include "Math/Integrator.h"

#include "Constants.h"

const double QEL_Mv = 0.71; // Vector mass in the dipole form factor,
↪   GeV^2
const double QEL_Ma = 1.0; // Axial vector mass in the dipole form
↪   factor, GeV^2
const double QEL_Fa0 = -1.2670; // Axial vector constant

double Ev = 1.0; // Global variable, neutrino energy, GeV

double ComputeCCQECrossSection(double q2_negative) {

    // Kinematic variables, parameters
    double neutrinoEnergy = Ev; // GeV
    double neutrinoEnergy2 = neutrinoEnergy * neutrinoEnergy;
    double q2 = -q2_negative; // Convention: q^2 = -Q^2
    double nucleonMass = mNeutron;
    double nucleonMass2 = TMath::Power(nucleonMass, 2);
    double nucleonMass4 = TMath::Power(nucleonMass2, 2);
    double GF2 = TMath::Power(GF, 2);
    double cosCabibbo2 = TMath::Power(TMath::Cos(CabibboAngle), 2);
    double leptonMass2 = TMath::Power(mMuon, 2);

    // Neutrino/antineutrino flag
    bool isNeutrino = true;
    int helicitySign = (isNeutrino) ? -1 : 1;

    // Dipole form factors
    double dipoleVector = 1.0 / TMath::Power(1 - q2 / QEL_Mv, 2);
    double electricProton = dipoleVector;
    double electricNeutron = 0.0;
    double magneticProton = MuP * dipoleVector;
    double magneticNeutron = MuN * dipoleVector;
```

```cpp
  // Sachs form factors
    double vectorElectric = electricProton - electricNeutron;
    double vectorMagnetic = magneticProton - magneticNeutron;


  // Dirac and Pauli form factors
    double F1V = (vectorElectric - q2 / (4 * nucleonMass2) *
↪ vectorMagnetic) / (1 - q2 / (4 * nucleonMass2));
    double F2V = (vectorMagnetic - vectorElectric) / (1 - q2 / (4 *
↪ nucleonMass2));


  // Axial form factors
    double axialFormFactor = QEL_Fa0 / TMath::Power(1 - q2 / QEL_Ma, 2);
    double inducedPseudoscalar = 2.0 * nucleonMass2 * axialFormFactor /
↪ (TMath::Power(mPion, 2) - q2);


  // Squared form factors
    double FA2 = TMath::Power(axialFormFactor, 2);
    double Fp2 = TMath::Power(inducedPseudoscalar, 2);
    double F1V2 = TMath::Power(F1V, 2);
    double F2V2 = TMath::Power(F2V, 2);


  // Kinematic variables
    double su = 4.0 * neutrinoEnergy * nucleonMass + q2 - leptonMass2;
    double q2_over_M2 = q2 / nucleonMass2;


  // Llewellyn Smith terms
    double termA = 0.25 * (leptonMass2 - q2) / nucleonMass2 * (
        (4 - q2_over_M2) * FA2 -
        (4 + q2_over_M2) * F1V2 -
        q2_over_M2 * F2V2 * (1 + 0.25 * q2_over_M2) -
        4 * q2_over_M2 * F1V * F2V -
        (leptonMass2 / nucleonMass2) * (
            F1V2 + F2V2 + 2 * F1V * F2V +
            FA2 + 4 * Fp2 + 4 * axialFormFactor * inducedPseudoscalar +
            (q2_over_M2 - 4) * Fp2
        )
    );

    double termB = -q2_over_M2 * axialFormFactor * (F1V + F2V);
    double termC = 0.25 * (FA2 + F1V2 - 0.25 * q2_over_M2 * F2V2);


  // Overall prefactor
    double prefactor = nucleonMass2 * GF2 * cosCabibbo2 / (8.0 * kPi *
↪ neutrinoEnergy2);
```

```
    // Total differential cross-section (in GeV^{-2})
    double xsec = prefactor * (termA + helicitySign * termB * su /
↪  nucleonMass2 + termC * TMath::Power(su, 2) / nucleonMass4);

    return xsec;
}
```

Since the cross section is differential with respect to a single variable, Q^2, it can be integrated numerically using standard techniques, such as those employed in previous examples using GSL.

However, it is important to remember that Q^2 is related to the Lorentz-invariant Mandelstam variable $s = 2E_\nu m_n + m_n^2$, where m_n is the neutron mass. For a fixed neutrino energy, s is fixed, and the kinematic limits of Q^2 follow from relativistic two-body scattering, see Eq. 5.9. These limits must be computed dynamically for each value of E_ν to ensure the integration is restricted to the physically allowed region. The total cross section divided by the incoming (anti)neutrino energy, σ/E_ν, is shown as a function of energy in Figs. 5.3 and 5.4.

Writing an event generator based on the Llewellyn Smith CCQE differential cross section follows a similar procedure to previous implementations. The phase-space kinematics are generated using the TGenPhaseSpace class (see Code Listing 3.5), and events are accepted/rejected based on the rejection sampling algorithm using the value of the differential cross-section function evaluated at each randomly generated phase-space point. It is important to note, however, that

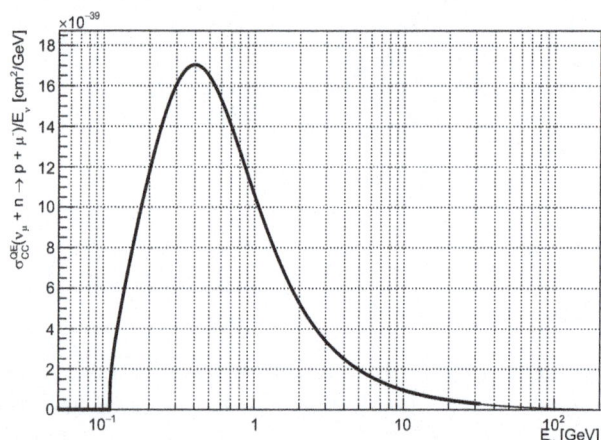

Fig. 5.3 The Llewellyn Smith charge-current quasi-elastic (CCQE) cross section divided by the neutrino energy for the process $\nu_\mu + n \rightarrow p + \mu^-$, for a free neutron target. The following parameter values were assumed in the model: $M_V^2 = 0.71 \text{ GeV}^2$ and $M_A^2 = 1.0 \text{ GeV}^2$

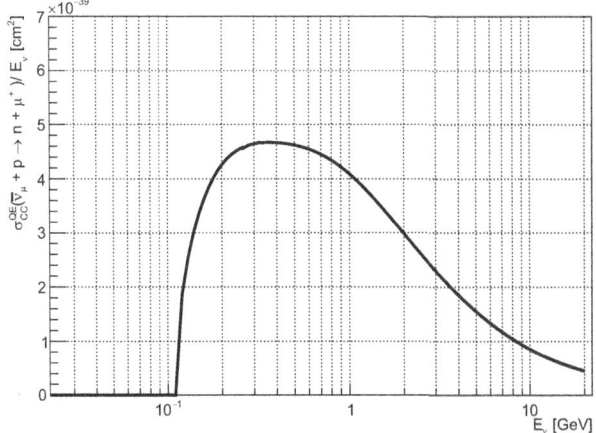

Fig. 5.4 The Llewellyn Smith charge-current quasi-elastic (CCQE) cross section divided by the neutrino energy for the process $\bar{\nu}_\mu + p \rightarrow n + \mu^+$, for a free proton target. The following parameter values were assumed in the model: $M_V^2 = 0.71 \, \text{GeV}^2$ and $M_A^2 = 1.0 \, \text{GeV}^2$

the Llewellyn Smith CCQE cross section is defined for a target nucleon at rest. Consequently, when accounting for the motion of nucleons inside a nucleus, one must boost to the rest frame of the struck nucleon before evaluating the cross section. Some examples of this procedure will be shown in the following sections.

5.2.6   Reconstruction of the Neutrino Energy

Let us now use relativistic kinematics to reconstruct the energy of the incoming neutrino from the final-state kinematics. Consider the charged-current quasi-elastic scattering process $\nu_l + n \rightarrow l^- + p$, where l is a charged lepton, n is a neutron initially at rest, and p is the outgoing proton. In four-momentum notation, the initial- and final-state vectors are given by

$$p_\nu = \begin{pmatrix} E_\nu \\ \mathbf{p}_\nu \end{pmatrix} ; \; p_n = \begin{pmatrix} M_n \\ \mathbf{0} \end{pmatrix} ; \; p_l = \begin{pmatrix} E_l \\ \mathbf{p}_l \end{pmatrix} ; \; p_p = \begin{pmatrix} E_p \\ \mathbf{p}_p \end{pmatrix} . \tag{5.39}$$

Squaring the proton four-momentum using energy-momentum conservation yields

$$p_p^2 = (p_\nu + p_n - p_l)^2 . \tag{5.40}$$

We have $p_p^2 = M_p^2$, and we can expand the right-hand side:

$$M_p^2 = p_\nu^2 + p_n^2 + p_l^2 + 2(p_\nu p_n) - 2(p_\nu p_l) - 2(p_n p_l) . \tag{5.41}$$

Assuming $p_\nu^2 = m_\nu^2 \simeq 0$, $|\mathbf{p}_\nu|^2 = E_\nu^2$, and $p_l^2 = m_l^2$, the scalar products become

$$p_\nu p_n = E_\nu M_n, \qquad p_\nu p_l = E_\nu E_l - E_\nu |\mathbf{p}_l| \cos\theta_l, \qquad p_n p_l = M_n E_l.$$

Substituting these into the previous equation gives

$$M_p^2 = M_n^2 + m_l^2 + 2E_\nu M_n - 2E_\nu (E_l - |\mathbf{p}_l| \cos\theta_l) - 2M_n E_l.$$

Rearranging terms:

$$2E_\nu (E_l - |\mathbf{p}_l| \cos\theta_l - M_n) = M_n^2 + m_l^2 - M_p^2 - 2M_n E_l.$$

Solving for the incoming neutrino energy:

$$E_\nu = \frac{M_n^2 + m_l^2 - M_p^2 - 2M_n E_l}{2(E_l - |\mathbf{p}_l| \cos\theta_l - M_n)}. \tag{5.42}$$

This expression depends only on known particle masses and final-state observables: the lepton energy, momentum, and scattering angle. However, because the struck neutron is bound inside a nucleus, it is effectively off-shell. To account for nuclear binding effects, the neutron mass is replaced by $M_n \to M_n - E_b$, where E_b is the binding energy. This leads to the modified reconstruction formula:

$$E_\nu = \frac{(M_n - E_b)^2 + m_l^2 - M_p^2 - 2(M_n - E_b)E_l}{2(E_l - |\mathbf{p}_l| \cos\theta_l - (M_n - E_b))}. \tag{5.43}$$

Using the phase-space sampling and rejection sampling techniques described earlier, and assuming a fixed neutrino energy of $E_\nu = 0.6\,\text{GeV}$ and a neutron target at rest, we generate a few thousand CCQE events and plot the various final-state kinematic distributions, see Fig. 5.5.

What observations can we make from the distributions shown in Fig. 5.5? Let us highlight a few key points:

- *Physical boundaries in distributions*: We can verify whether the physical limits derived earlier from relativistic kinematics are respected. For instance, the final-state muon energy falls within the expected boundaries from Eq. 5.13. Similarly, the distribution of the negative squared momentum transfer, Q^2, adheres to the expected range, from $Q_{\min}^2 = 0$ to $Q_{\max}^2 = (s - M_n^2)(s - M_p^2)/s \simeq 0.63\,\text{GeV}^2$.
- *Momentum transfer*: Recall that the squared momentum transfer q^2 quantifies the momentum imparted from the incoming neutrino to the nucleon. When the transfer is maximal, the muon momentum is minimal, and vice versa. However, in the plots we use the conventional definition $Q^2 = -q^2$, so the relationship is inverted: Small values of Q^2 correspond to high muon energies. This correlation between Q^2 and the final-state muon kinematics is also evident in the distribution

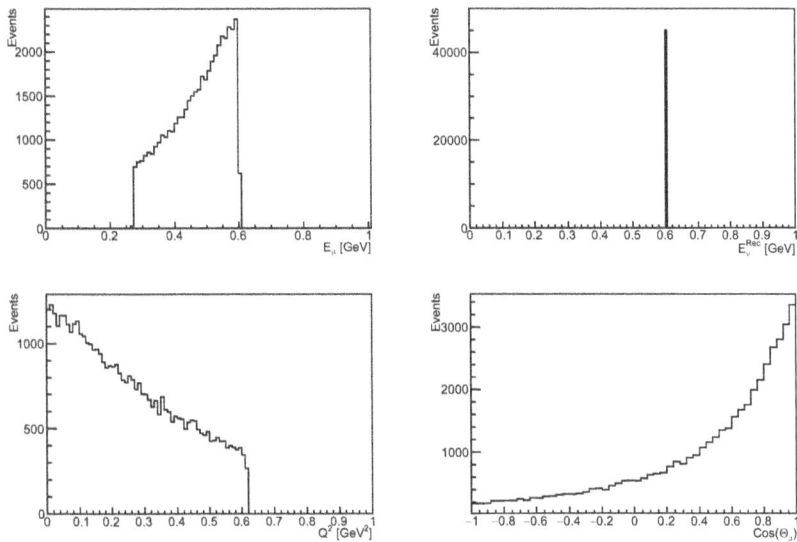

Fig. 5.5 Final-state kinematics distribution for the CCQE process, $\nu_\mu + n \rightarrow \mu + p$, using the Llewellyn Smith differential cross section, and assuming a fixed $E_\nu = 0.6$ GeV energy and a neutron target at rest

of Q^2 versus the cosine of the muon scattering angle, $\cos\theta_\mu$, shown in Fig. 5.6. As expected from Eq. 5.7, forward-scattered muons (large $\cos\theta_\mu$) correspond to minimal Q^2 values.

- *Reconstructed neutrino energy*: The expression in Eq. 5.43 successfully recovers the true neutrino energy, $E_\nu = 0.6$ GeV. In this specific case, the target neutron is free and at rest, so the correction due to the nucleon binding energy is not required.

Although the case of a free, stationary nucleon is conceptually simple, it is far from realistic. In practice, nucleons are bound within nuclei and possess intrinsic momentum due to Fermi motion. In the following, we slightly extend the model to incorporate some version of this effect by assigning a random initial momentum and binding energy to the target nucleon. This leads to a broadening of the kinematic distributions and introduces several additional model parameters and associated uncertainties.

5.3   Quasi-elastic Scattering on Bound Nucleons in Motion

We now give a brief overview of how the motion and binding of nucleons inside nuclei are treated in neutrino event generators and outline the basic elements needed to go beyond the idealized picture of a static, free nucleon used thus far in our discussion of neutrino-nucleon scattering.

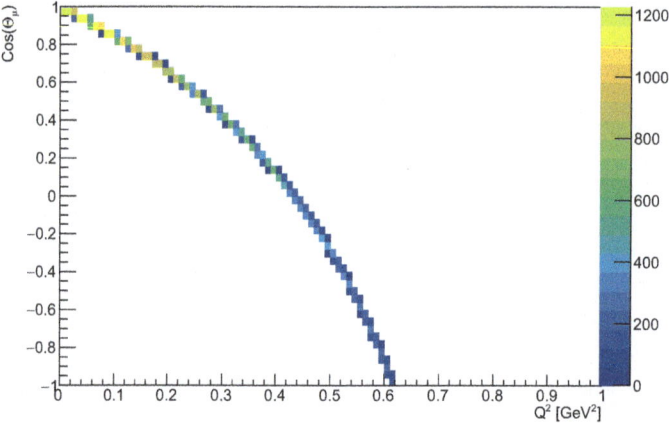

Fig. 5.6 Negative momentum transfer squared, Q^2, against the cosine of the final-state lepton angle, $\cos \Theta_\mu$, for the CCQE process, $\nu_\mu + n \rightarrow \mu + p$, using the Llewellyn Smith differential cross section, and assuming a fixed $E_\nu = 0.6$ GeV energy and a neutron target at rest

Nucleons are bound within atomic nuclei by the strong nuclear force. Their states are characterized by spatial density distributions, momentum distributions, and binding energies. In neutrino event generators, the density profile is typically treated independently from the momentum and binding energy. Two commonly used models for describing nucleon states are the *global relativistic Fermi gas* (RFG) model and the *local Fermi gas* (LFG) model. In the global model, nucleons are treated as moving in a single, uniform potential well, whereas in the local model, nucleons occupy position-dependent potential wells (typically spherical shells) defined by the Woods-Saxon[9] potential [11]:

$$V(r) = -\frac{V_0}{1 + \exp[(r - R)/a]}, \tag{5.44}$$

where V_0 is the potential depth, r is the radial position, R is the nuclear radius, and a is the surface thickness parameter, which controls the diffuseness of the nucleon distribution at the edge of the nucleus. The parameters of the Woods-Saxon potential are usually fitted to data from elastic nucleon-nucleus scattering experiments. Both the global and local Fermi gas models treat nucleons as noninteracting fermions.

A key feature of nucleons bound in a nucleus is their motion, often described by the *Fermi momentum*, p_F, which typically lies in the range 200–300 MeV/c. This quantity characterizes the maximum momentum of nucleons due to the Pauli exclusion principle. Several models exist to describe this motion, incorporating nuclear structure, quantum statistics, and interaction dynamics. Functions that

---

[9] Named after American physicists R. D. Woods and D. S. Saxon (1914–2005).

jointly describe the momentum and binding energy distributions of nucleons are referred to as *spectral functions*. Spectral function models may be one dimensional (momentum dependence only) or two dimensional (momentum and binding/removal energy dependence).

In what follows, we adopt a one-dimensional spectral function example based on the global relativistic Fermi gas model [12]. Other notable models for quasi-elastic (1p1h) scattering include the Benhar spectral function[10] [13] and the Smith-Moniz model[11] [14]. Going beyond the quasi-elastic picture, models developed by Nieves *et al.* [15] include multinucleon interactions (2p2h) and collective nuclear effects.

Binding energies also play an important role in describing bound nucleon systems. For example, the deuteron—the simplest bound nucleon system—has a binding energy of $E_b \simeq 2.2$ MeV. The nuclear binding energy per nucleon increases rapidly with atomic number, reaching a maximum near $_{28}^{62}$Ni with approximately 8.8 MeV per nucleon, after which it decreases slowly for heavier nuclei. Excluding very light nuclei, the average binding energy is approximately constant, around 8 MeV per nucleon. This quantity is crucial in neutrino experiments, as the initial neutrino energy must be inferred from final-state kinematics in scattering on bound nucleons. Uncertainties or biases in the binding energy can therefore propagate to systematic uncertainties in neutrino energy reconstruction and ultimately affect the determination of neutrino oscillation parameters.[12]

Closely connected to Fermi motion is the phenomenon of *Pauli blocking*. Because nucleons are fermions, they occupy discrete quantum states up to a maximum energy—the *Fermi energy*. After a neutrino-nucleon interaction, the final-state nucleon must transition into an unoccupied state above the Fermi surface. This restriction in available phase space suppresses certain interaction kinematics and is referred to as Pauli blocking. It represents a direct consequence of the quantum-statistical nature of nucleons inside the nucleus.

The simplest model of Fermi motion is commonly referred to as the *pure Fermi gas* model. In this approach, the nuclear ground state is treated as a superposition of noninteracting, degenerate gases of neutrons and protons, each characterized by momentum distributions $n_n(p)$ and $n_p(p)$, respectively. For a pure (zero-temperature) Fermi gas, the occupation number is given by

$$n_i(p) = \theta(p_F - |\mathbf{p}|), \tag{5.45}$$

where θ is the Heaviside[13] step function, p_F is the Fermi momentum, and i labels the nucleon species ($i = n$ for neutrons and $i = p$ for protons). This

---

[10] Named after Italian theoretical physicist O. Benhar (1953-).

[11] Named after American theoretical physicists R. A. Smith and E. J. Moniz (1944-).

[12] In addition to binding energy, other related quantities are also used in modelling, especially in the context of electron scattering data. An important example is the *removal energy*, which appears in spectral function models and quasi-elastic scattering analyses [16].

[13] Named after British physicist and mathematician O. Heaviside (1850–1925).

form corresponds to a step-function occupation of momentum states: All states with $|\mathbf{p}| < p_F$ are filled, and all others are empty. At nonzero temperature, the occupation could instead follow the Fermi-Dirac distribution. When normalized, $n_i(p)$ represents the probability that a state with momentum p is occupied in the initial configuration. For a final state to be available after interaction, the occupation factor must satisfy $1 - n_i(p) > 0$, ensuring the Pauli exclusion principle is respected.

Historically, Fermi gas momentum distributions have been extracted from quasi-elastic electron scattering data on heavy nuclei. Before specifying the details of the model further, let us estimate the average kinetic energy of a nucleon within the nucleus, under a few simplifying assumptions. We assume an equal number of protons and neutrons, a continuous distribution of momentum states, and complete filling of all states up to the Fermi momentum. Then, for a nucleus containing A nucleons confined to a volume Ω, the total number of occupied states is

$$A = \frac{4\Omega}{h^3} \int_0^{p_F} d^3p = \frac{16\pi}{3} \Omega \left(\frac{p_F}{h}\right)^3, \qquad (5.46)$$

where h is Planck's constant, and h^3 represents the volume of a single quantum state in phase space. The prefactor 4 accounts for the four available spin-isospin combinations (two spin states and two isospin states) per momentum mode. The total kinetic energy of the nucleons in this model is given by

$$T = \frac{4\Omega}{h^3} \int_0^{p_F} \frac{p^2}{2m_0} d^3p = \frac{16\pi}{h^3} \Omega \cdot \frac{1}{2m_0} \cdot \frac{p_F^5}{5} = \frac{3}{5} \frac{p_F^2}{2m_0} A, \qquad (5.47)$$

where m_0 is the rest mass of the nucleon. Thus, the average kinetic energy per nucleon is

$$\langle T \rangle = \frac{T}{A} = \frac{3}{5} \frac{p_F^2}{2m_0}. \qquad (5.48)$$

As an example, in the GENIE neutrino event generator [1], the Fermi momenta for argon are $p_{F,n} = 259$ MeV/c for neutrons and $p_{F,p} = 242$ MeV/c for protons. These values correspond to an average kinetic energy per nucleon of approximately $T/A \simeq 20$ MeV.

In our implementation, we adopt a parametrization after Bodek and Ritchie [12] to model Fermi motion. This approach is available in neutrino event generators and is based on fits to data from electron scattering experiments on heavy nuclei. The model extends the pure Fermi gas model by relaxing the strict cutoff at the Fermi momentum. It introduces a high-momentum tail in the nucleon momentum distribution, motivated by nucleon-nucleon correlations within the nuclear medium. The normalized momentum distribution is defined such that, for a given momentum $\mathbf{p}$, the probability density is

$$\frac{dP}{dp} = |\phi(\mathbf{p})|^2 \, 4\pi |\mathbf{p}|^2, \tag{5.49}$$

where the momentum-space wave function squared, $|\phi(\mathbf{p})|^2$, is given by

$$|\phi(\mathbf{p})|^2 = \begin{cases} \frac{1}{C}\left[1 - 6\left(\frac{p_F a}{\pi}\right)^2\right] & \text{for } 0 < |\mathbf{p}| < p_F, \\ \frac{1}{C}\left[2R\left(\frac{p_F a}{\pi}\right)^2\left(\frac{p_F}{|\mathbf{p}|}\right)^4\right] & \text{for } p_F < |\mathbf{p}| < 4 \text{ GeV}/c, \\ 0 & \text{for } |\mathbf{p}| > 4 \text{ GeV}/c. \end{cases}$$

Here, $a = 2 \text{ GeV}^{-1}/c$, $C = \frac{4}{3}\pi p_F^3$ is the normalization constant, and $R = [1 - p_F/(4 \text{ GeV}/c)]^{-1}$. The model accounts for differences in the Fermi momenta of protons and neutrons through the scaling relations:

$$p_{F,p} = p_F \left(\frac{2Z}{A}\right)^{1/3}, \qquad p_{F,n} = p_F \left(\frac{2N}{A}\right)^{1/3}, \tag{5.50}$$

where Z and N are the proton and neutron numbers, respectively, and $A = Z + N$ is the mass number. The reference Fermi momentum p_F corresponds to that of an isoscalar target ($Z = N = A/2$). The model requires input values for the Fermi momentum p_F, which vary by nucleus. Typical values used in the GENIE generator [1] are summarized in Table 5.3, and an example momentum distribution for neutrons in ^{12}C is shown in Fig. 5.7.

We now implement the global relativistic Fermi gas model with modifications according to the specified model for the case of a neutron inside $^{12}_6$C. We begin by constructing a one-dimensional histogram representing the probability distribution of the nucleon's momentum magnitude, based on the model parametrization. We then sample from this histogram to generate random nucleon momentum vectors. These vectors are interpreted in the laboratory frame, where the nucleus is at rest, and are converted into four-momentum vectors assuming the mass of a free neutron.

Table 5.3 Fermi momentum and nuclear removing energy values used for some isotopes [1].

Isotope	Fermi momentum (neutron, proton) [MeV/c]	Nuclear removing energy [MeV]
^{6}Li	169,169	17.0
^{12}C	221,221	25.0
^{16}O	225,225	27.0
^{40}Ar	259,242	29.5
^{56}Fe	263,251	36.0
^{208}Pb	283,245	44.0

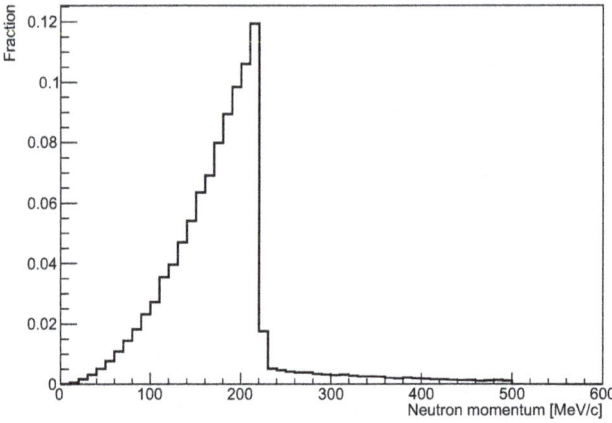

Fig. 5.7 Neutron momentum distribution in the laboratory frame for $^{12}_{6}$C according to the global relativistic Fermi gas model with Bodek-Ritchie modification

Code listing 5.2 Code listing for the global relativistic Fermi gas model following Bodek and Ritchie [12]

```
#include <TMath.h>
#include <TH1D.h>
#include <TRandom2.h>
#include <TLorentzVector.h>
#include <TVector3.h>
#include <iostream>

#include "Constants.h"

TRandom2* rng = new TRandom2();

// Generate Bodek-Ritchie probability distribution for a neutron in 12C
TH1D* createBodekRitchieDistribution() {
  const double maxMomentum = 4.0; // [GeV/c], cutoff of distribution
  const double highTailLimit = 0.5; // [GeV/c], upper bound of tail
↪  region
  const double atomicMass = 12.0;
  const double atomicNumber = 6.0;
  const double neutronCount = atomicMass - atomicNumber;
  const double baseFermiMomentum = 0.221; // [GeV/c], isoscalar Fermi
↪  momentum

  // Fermi momentum scaled for neutron in 12C
  const double fermiMomentum = baseFermiMomentum * TMath::Power(2.0 *
```

```
  neutronCount / atomicMass, 1.0 / 3.0);

  // Bodek-Ritchie model parameters
  const double aParam = 2.0; // [GeV/c]^{-1}
  const double normalizationConst = 4.0 * kPi *
↪   TMath::Power(fermiMomentum, 3) / 3.0;
  const double R = 1.0 / (1.0 - fermiMomentum / maxMomentum);

  // Create probability histogram
  const unsigned int numBins = static_cast<unsigned int>(1000 *
↪   maxMomentum);
  const double binWidth = maxMomentum / (numBins - 1);
  TH1D* momentumDist = new TH1D("", "", numBins, 0, maxMomentum);

  for (unsigned int i = 0; i < numBins; ++i) {
    const double momentum = i * binWidth;
    const double momentumSquared = momentum * momentum;

    // Momentum-space wavefunction squared
    double phiSquared = 0.0;
    if (momentum <= fermiMomentum) {
      phiSquared = (1.0 / normalizationConst) * (1.0 - 6.0 *
↪   TMath::Power(fermiMomentum * aParam / kPi, 2));
    } else if (momentum > fermiMomentum && momentum < highTailLimit) {
      phiSquared = (1.0 / normalizationConst) *
                    (2.0 * R * TMath::Power(fermiMomentum * aParam / kPi,
↪   2) *
                    TMath::Power(fermiMomentum / momentum, 4));
    }

    // Probability density dP/dp = |phi(p)|^2 x 4 * pi*p^2
    const double probDensity = 4.0 * kPi * momentumSquared * phiSquared;
    momentumDist->Fill(momentum, probDensity);
  }

  // Normalize the histogram to unit area
  momentumDist->Scale(1.0 / momentumDist->Integral("width"));

  return momentumDist;
}

// Generate a random nucleon 4-momentum from the Bodek-Ritchie
↪   distribution
TLorentzVector generateNucleon4Vector(TH1D* momentumDist) {
```

```cpp
  if (!momentumDist) {
    std::cerr << "Error: Probability distribution not provided." <<
↪ std::endl;
    std::exit(1);
  }


  const double sampledMomentum = momentumDist->GetRandom();


  // Generate random direction in spherical coordinates
  const double cosTheta = -1.0 + 2.0 * rng->Rndm();
  const double sinTheta = TMath::Sqrt(1.0 - cosTheta * cosTheta);
  const double phi = 2.0 * kPi * rng->Rndm();


  const double px = sampledMomentum * sinTheta * TMath::Cos(phi);
  const double py = sampledMomentum * sinTheta * TMath::Sin(phi);
  const double pz = sampledMomentum * cosTheta;


  TVector3 momentumVec(px, py, pz);


  TLorentzVector nucleon;
  nucleon.SetVect(momentumVec);
  nucleon.SetE(TMath::Sqrt(momentumVec.Mag2() + mNeutron * mNeutron));


  return nucleon;
}


// Generate nucleon momenta from the BR-modified Fermi gas model
void simulateBodekRitchieFermiGas() {
  TH1D* fermiMomentumDistribution = createBodekRitchieDistribution();
  TH1D* histogram = new TH1D("GFGmom", "", 60, 0, 600); // Histogram in
↪ MeV/c


  const int numEvents = 10000;


  for (int i = 0; i < numEvents; ++i) {
    TLorentzVector nucleon4Vec =
↪ generateNucleon4Vector(fermiMomentumDistribution);
    histogram->Fill(nucleon4Vec.Vect().Mag() * 1000.0); // Convert to
↪ MeV/c
  }


  histogram->Scale(1.0 / histogram->Integral());
  histogram->Draw("hist");
}
```

The momentum distribution obtained from the code is shown in Fig. 5.7.

However, using a momentum distribution alone is not sufficient to fully describe interactions involving moving, bound nucleons. When simulating neutrino-nucleon interactions that include Fermi motion, it is essential to account for the fact that the interacting nucleon is initially in an off-shell bound state within the nucleus. Consider, for example, the charged-current quasi-elastic process $\nu_l + A \rightarrow (A - 1) + p + l$, where A is the initial nucleus, and $(A - 1)$ denotes the residual spectator system.

In such cases, one common approach is to estimate the energy of the off-shell nucleon by considering the energy balance between the initial and final nuclear states. The remnant $(A - 1)$ nucleus will have a mass reduced by the missing nucleon, accounting also for the binding (or removal) energy. This is expressed as

$$M' = M - (M_n - E_b), \tag{5.51}$$

where M is the mass of the initial nucleus, M' is the mass of the residual $(A - 1)$ system, M_n is the mass of the removed nucleon, and E_b is its binding energy.

The lab-frame energy of the bound nucleon, E_n, can then be estimated from energy conservation as the difference between the energy of the initial nucleus at rest and the energy of the recoiling remnant:

$$E_n = M - \sqrt{M'^2 + \mathbf{p}_n^2}, \tag{5.52}$$

where $\mathbf{p}_n$ is the momentum of the interacting nucleon.[14]

5.3.1 Reference Frames

When simulating the phase-space kinematics of neutrinos scattering on moving, bound nucleons in the laboratory frame, it is necessary to perform Lorentz boosts to other frames where calculations are more conveniently or conventionally performed. In particular, the charged-current quasi-elastic (CCQE) differential cross section (in the Llewellyn Smith formalism) is defined in the rest frame of the nucleon. This leads us to consider three distinct reference frames relevant for the simulation:

- *Laboratory frame*: This is the reference frame in which the simulation is typically performed. The incoming neutrino four-momentum is defined in this frame, and the target nucleus is at rest. The nucleon, however, is in motion due to Fermi motion. The nucleus four-momentum is $p_N = (0, 0, 0, M_N)$, where M_N is the nuclear mass.

---

[14] It is important to emphasize that this is a simplified model used to illustrate the essential features of the problem. The full treatment of nuclear kinematics is considerably more complex. For a detailed discussion, see Ref. [16].

- *Nucleon rest frame*: This is the frame in which the Llewellyn Smith CCQE differential cross section is evaluated. The interacting nucleon is at rest, $p_n = (0, 0, 0, M_n)$, where M_n is the nucleon mass. Both the neutrino four-momentum and the outgoing lepton/nucleon kinematics must be boosted into this frame for proper evaluation.
- *Centre-of-mass (CM) frame of the neutrino + nucleon system*: In this frame, the incoming neutrino and initial nucleon momenta are equal and opposite, and the final-state lepton and nucleon are emitted back to back. This frame is often useful for computing final-state momenta analytically and then boosting back to the lab frame.

Having identified these reference frames and discussed the treatment of the off-shell initial nucleon, we can now illustrate the process with a code example. Code listing 5.3 shows how to generate a bound nucleon with Fermi motion according to the Fermi gas model with Bodek-Ritchie modifications, and how to compute the necessary boost vectors to move between the laboratory, nucleon rest, and CM frames.

The procedure begins by defining the incoming neutrino beam energy and the nucleon binding energy for $^{12}_{6}C$, along with other relevant nuclear properties. A four-momentum vector is then sampled from the global relativistic Fermi gas momentum distribution. The off-shell energy of the nucleon is computed according to the simplified model discussed earlier, and the resulting four-vector represents the initial state of the interacting nucleon. Finally, Lorentz boosts are constructed to transform into the nucleon rest frame or the beam-nucleon centre-of-mass frame for further kinematic evaluation.

Code listing 5.3 Code listing for an initial nucleon state preparation with various reference frames involved

```
...
// Binding/removal energy for 12C
double Eb = 0.0250;// GeV
// Neutrino beam energy in Lab frame
double Ebeam = 0.6; // GeV

double C_A = 12; // Atomic mass number, A
unsigned int C_Z = 6; // Atomic number, Z
double MC = C_A*au; // Mass, GeV

Int_t Nevents = 2;
for (Int_t nEv=0;nEv<Nevents;nEv++) {

    // Neutrino beam defined in the L A B  F R A M E
    TLorentzVector beam(0.0,0.0, Ebeam, Ebeam);
```

```cpp
//----------------------------------------------------------
// Set the initial nucleon kinematics

// Generate the initial nucleon 4-momentum in the Lab frame
TLorentzVector p4Nucleon = generateNucleon4Vector(probDistBR);

// The 3-momentum vector
TVector3 p3Nucleon = p4Nucleon.Vect();
GFGmom->Fill(p3Nucleon.Mag()*1000); // [MeV/c]

// Initial nucleus mass for 12C
double Mini = MC;
// Final nucleon mass after removing a neutron
double Mfin = Mini - (mNeutron - Eb);

// The Lab-frame off-shell initial nucleon energy
double Enini = Mini - std::sqrt( Mfin*Mfin + p3Nucleon.Mag2() );

// Set the Lab-frame nucleon 4-momentum for the interaction
p4Nucleon.SetVect( p3Nucleon );
p4Nucleon.SetE( Enini );

//----------------------------------------------------------
// Get the Lorentz boost vectors

 // L A B   F R A M E   <--> N U C L E O N   A T   R E S T   F R A M E
TVector3 beta_lab_to_nRest = (p4Nucleon).BoostVector();
TVector3 beta_nRest_to_lab = -beta_lab_to_nRest;

// L A B   F R A M E   <--> ( B E A M + T A R G E T )   C O M   F R A M E
TVector3 beta_lab_to_beamtargetCOM = (p4Nucleon+beam).BoostVector();
TVector3 beta_beamtargetCOM_to_lab = -beta_lab_to_beamtargetCOM;

//----------------------------------------------------------
// Boost back from the Lab frame to the Nucleon at rest frame
// where we can evaluate the LL-Smith cross-section
TLorentzVector beam_nRest = TLorentzVector( beam );
TLorentzVector nucleon_nRest = TLorentzVector( p4Nucleon );
beam_nRest.Boost( -beta_lab_to_nRest );
nucleon_nRest.Boost( -beta_lab_to_nRest );

double Enu_Lab = beam.E();
double Enu = beam_nRest.E();
```

```
//------------------------------------------------------------
  // Sanity checks:
  std::cout << "nu 4-mom in Lab frame: " << std::endl;
  beam.Print();
  std::cout << "nu 4-mom in nucleon rest frame: " << std::endl;
  beam_nRest.Print();
  std::cout << "Nucleon 4 -mom in Lab frame: " << std::endl;
  p4Nucleon.Print();
  std::cout << "Nucleon 4-mom in nucleon rest frame: " << std::endl;
  nucleon_nRest.Print();
  std::cout << "s value in Lab frame: " << (beam+p4Nucleon).Mag2() <<
↪ std::endl;
  std::cout << "s value in nRest frame: " <<
↪ (beam_nRest+nucleon_nRest).Mag2() << std::endl;
  std::cout << "-------------------------------------------" <<
↪ std::endl;
  }
  ...
```

The code snippet includes several sanity checks to examine the four-momentum components in the different reference frames and to verify that the Mandelstam variable $s = (p_1 + p_2)^2$ remains invariant across frames, as it is a Lorentz-scalar. A representative example in code listing 5.4 illustrates this consistency.

In this example, the neutrino energy differs significantly between the laboratory frame and the rest frame of the nucleon. While in the lab frame the neutrino energy is set to $E_{\nu,\text{Lab}} = 0.6$ GeV, its energy in the nucleon rest frame is found to be approximately $E_{\nu,\text{nrest}} \simeq 0.97$ GeV after boosting. This reflects the fact that nucleons can carry sizable momenta due to Fermi motion—comparable in magnitude to the energies of the neutrino beam.

In the nucleon rest frame, the interacting nucleon has only a time-like component, as expected. However, its energy is found to be *less* than the free nucleon mass, consistent with the nucleon being off-shell due to nuclear binding. Despite these differences in individual energy components, the value of the Lorentz-invariant Mandelstam variable s is identical in both frames, confirming the consistency of the boost and event setup.

Code listing 5.4 Output from the code snippet showing the four-momenta from various reference frame

```
nu 4-mom in Lab frame:
(x,y,z,t)=(0.000000,0.000000,0.600000,0.600000)
nu 4-mom in nucleon rest frame:
(x,y,z,t)=(-0.350355,0.111312,0.895232,0.967770)
Nucleon 4 -mom in Lab frame:
```

```
(x,y,z,t)=(0.369918,-0.117527,-0.311717,0.902512)
Nucleon 4-mom in nucleon rest frame:
(x,y,z,t)=(-0.000000,-0.000000,0.000000,0.752800)
s value in Lab frame: 2.02378
s value in nRest frame: 2.02378
```

After this detailed preparation, the phase-space kinematics for the interaction can be generated using the standard TGenPhaseSpace class (or other similar tools). The Llewellyn Smith CCQE differential cross section can then be applied within a rejection sampling algorithm, provided it is evaluated in the rest frame of the nucleon. Care must be taken to correctly boost four-momentum vectors between the relevant reference frames at each step of the simulation.

One important consequence of this modelling is that the reconstructed neutrino energy, obtained from final-state observables in the laboratory frame (such as the lepton kinematics and the known rest masses of the nucleon and lepton), becomes smeared. This smearing arises from the random motion of the initial-state nucleons due to Fermi motion and is illustrated in Fig. 5.8. Therefore, the nucleon's motion in the nucleus imposes a fundamental limitation on the neutrino energy reconstruction.

It is also important to emphasize that the discussion so far assumes a pure sample of CCQE scattering events. In practice, this assumption rarely holds: Various other processes—such as inelastic scattering and multinucleon interactions—can mimic CCQE-like final states and further distort the reconstructed neutrino energy. In the following, we consider one such process (albeit briefly): single-pion production via baryon resonance excitation.

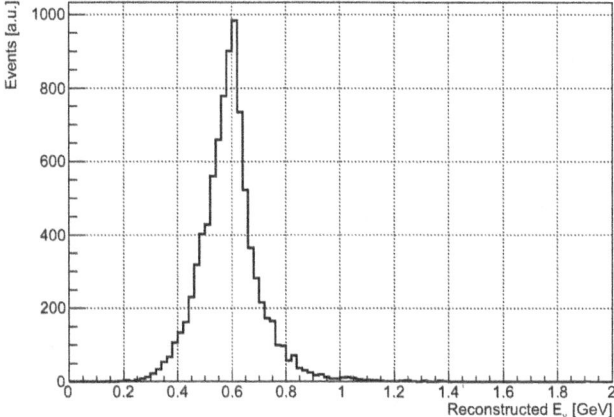

Fig. 5.8 Reconstructed neutrino energy from generated CCQE events, $\nu_\mu + n \rightarrow p + \mu$, with $E_\nu = 0.6$ GeV, using the global relativistic Fermi gas model with Bodek-Ritchie modification for the initial state nucleon motion

5.4   Single-Pion Resonance

Nucleons inside nuclei can be excited to higher-energy states due to their internal structure as bound states of three strongly interacting quarks. These excitations manifest as *resonances*. Resonant states are typically classified by their orbital angular momentum L, isospin $2I$, and spin $2S$ and are conventionally labelled as $L_{2I, 2S}$. The possible combinations of quantum numbers can be derived using isospin algebra.

A *single-pion resonance* process refers to an intermediate excited nucleon state that subsequently decays into a final-state nucleon and a pion, for example, (π^+, p). Because isospin is approximately conserved in strong interactions, the isospin structure of the initial and final states plays an important role in labelling and calculating the allowed transitions. The three pions—π^+, π^0, and π^-—form an isospin triplet with total isospin $I = 1$ and third-component values $I_3 = +1, 0, -1$, respectively. Protons and neutrons form an isospin doublet with $I = 1/2$, with $I_3 = +1/2$ for the proton, and $I_3 = -1/2$ for the neutron. The composite isospin states formed by pion-nucleon combinations, such as (π^+, p) or (π^0, n), can be expressed in terms of eigenstates of total isospin using Clebsch-Gordan[15] coefficients. Among these, the lowest-lying resonances (with total isospin $I = 3/2$ or $I = 1/2$ and small orbital angular momentum L) provide the dominant contributions to single-pion production at low energies.

There are many possible resonances, but in this section we focus on the most prominent example: the $P_{33}(1232)$ resonance, also known as the $\Delta(1232)$ baryon. This resonance has a very short lifetime and an almost 100% branching ratio to decay into a nucleon and a pion. Other resonances may have different decay channels and branching ratios, including decays to photons or multi-pion states.

The presence of a pion in the final state significantly alters the event kinematics. If the pion escapes detection, it can lead to a misidentification of the interaction as a CCQE-like event, biasing the reconstructed neutrino energy. In some cases, the pion produced by the Δ decay may be reabsorbed within the nuclear medium before leaving the nucleus, further complicating the identification of the interaction. Such effects must be modelled in neutrino event generators and are typically constrained using data and simulations. Here, we focus only on the basic aspects of implementing a single-pion resonance mechanism.

One can estimate the typical kinetic energy of a pion emerging from a $\Delta(1232)$ resonance decay. Consider the inverse process: a π^0 colliding with a stationary proton to form a Δ resonance, $\pi^0 + p \rightarrow \Delta$. Using four-momentum conservation:

$$p_\pi + p_p = p_\Delta$$

$$(p_\pi + p_p)^2 = p_\Delta^2 = m_\Delta^2$$

---

[15] Named after German mathematicians A. Clebsch (1833–1872) and P. Gordan (1837–1912).

$$m_\pi^2 + m_p^2 + 2m_p E_\pi = m_\Delta^2$$

$$E_\pi = \frac{m_\Delta^2 - m_p^2 - m_\pi^2}{2m_p}, \tag{5.53}$$

where E_π is the energy of the pion in the lab frame. The corresponding pion kinetic energy is

$$T_\pi = E_\pi - m_\pi = \frac{m_\Delta^2 - (m_p + m_\pi)^2}{2m_p} \simeq 190 \text{ MeV}, \tag{5.54}$$

assuming $m_\pi \simeq 140$ MeV, $m_p \simeq 938$ MeV, and $m_\Delta = 1232$ MeV. This provides an estimate of the typical energy scale of a pion from Δ resonance decay in a free nucleon system, $T_\pi \approx 200$ MeV.

In a nuclear environment, the situation is more complicated. The nucleon is bound and influenced by the nuclear potential, other nucleons may act as spectators, and the effective kinematics can be altered by many-body effects. The pion's energy may depend on the orbital angular momentum of the many-body nuclear state, and the pion itself may undergo final-state interactions or reabsorption. Nevertheless, this simple estimate shows that such pions can carry significant kinetic energy, and if undetected, they can strongly distort the inferred kinematics of the interaction. In addition, as we will see, the total production cross section for single-pion resonance processes is comparable in magnitude to that of the CCQE interaction.

5.4.1 The Rein-Sehgal Model

We now turn to the implementation of single-resonance production and decay in neutrino-nucleon scattering, based on the model developed by Rein and Sehgal [17]. While other models exist for describing resonance processes, our objective here is to capture the essential features using a well-established framework and to generate final-state kinematics for a nucleon at rest. The model can be extended to include Fermi motion by following steps similar to those described previously—namely, sampling nucleon momenta from a Fermi gas distribution and performing the appropriate Lorentz boosts—but to maintain simplicity, we focus here on the rest frame case.

A CC single-pion resonance process consists of two stages: the *production* of the resonance via a weak charged-current interaction (e.g. $\nu + n \rightarrow N^* + \mu$) and the subsequent *decay* of the resonance (e.g. $N^* \rightarrow n + \pi^+$). Each stage is described by a corresponding quantum mechanical amplitude. Because the intermediate resonance is not directly observed—and because multiple resonances may contribute and interfere—the observable final-state nucleon-pion system can be used to classify the contributing resonant states.

A given nucleon-pion final state, characterized by quantum numbers $L_{2I, 2S}$, can in principle arise from several intermediate resonances. For example, the final state

(n, π^+) corresponds to a combination of isospin states that could be either $I = 1/2$ or $I = 3/2$, based on the Clebsch-Gordan decomposition of the isospin components $I_3(\pi^+) = +1$ and $I_3(n) = -1/2$. If the final state corresponds to total isospin $I = 3/2$, then all resonances with $I = 3/2$ may contribute. In this case, the total amplitude for the charged-current (CC) or neutral-current (NC) process is obtained by summing the amplitudes of the contributing resonances:

$$A^{CC,NC} \propto \sum a^{CC,NC}(N^*), \tag{5.55}$$

where $a^{CC,NC}(N^*)$ denotes the amplitude for producing a given resonance N^* via a CC or NC process. Similarly, the amplitudes for final states with $I = 1/2$ must be summed over all contributing $I = 1/2$ resonances.

Each amplitude $a(N^*)$ can be factorized into two parts: a *production amplitude* describing the weak interaction that produces the resonance, and a *decay amplitude* describing the subsequent hadronic decay. This is often expressed as

$$a_k^{CC}(N^*) = f_k^{CC}(v + n \rightarrow N^*) \times \eta(N^* \rightarrow n + \pi^+), \tag{5.56}$$

where k labels the helicity of the production amplitude (associated with the W-boson polarization vector components in the model). The decay amplitude η includes a Breit-Wigner factor[16] describing the resonance mass distribution. The helicity amplitudes f_k in the model of Rein and Sehgal are derived from the relativistic quark model developed by Feynman, Kislinger, and Ravndal [18, 19], which is based on a symmetric harmonic oscillator description of quarks. For full technical details, we refer the reader to the original Rein-Sehgal paper. In what follows, we summarize the main components relevant to our chosen example: the $P_{33}(1232)$ resonance, also known as the $\Delta(1232)$.

The general differential cross section for single-pion resonance production in neutrino-nucleon scattering, assuming a nucleon at rest, is given by the Rein-Sehgal model [17] as

$$\frac{d\sigma}{dQ^2\,dW} = \frac{G_F^2}{4\pi^2}\left(\frac{-q^2}{Q^2}\right)\frac{W}{M}\kappa\left[u^2\sigma_L + v^2\sigma_R + 2uv\sigma_S\right], \tag{5.57}$$

where M is the nucleon mass, W the invariant mass of the hadronic system, and in this frame the four-momentum transfer to the hadronic system is $q = (v, \mathbf{Q})$ with v the energy transfer to the nucleon and $\mathbf{Q}$ the three-momentum transfer, $q^2 = v^2 - Q^2$. The cross section is given in terms of the different polarization states of the intermediate vector boson. The terms $\sigma_{L,R,S}$ are the partial cross sections for the absorption of the W-boson with left-handed, right-handed, or scalar (longitudinal)

---

[16] Named after American physicist G. Breit (1899–1981) and Hungarian-American physicist E. P. Wigner (1902–1995).

helicity, respectively. These contribute incoherently and include the resonance decay amplitude. For antineutrino scattering, the roles of σ_L and σ_R are interchanged.

The kinematic factors u, v, and κ are defined as

$$u = \frac{E + E' + Q}{2E}, \tag{5.58}$$

$$v = \frac{E + E' - Q}{2E}, \tag{5.59}$$

$$\kappa = \frac{W^2 - M^2}{2M}, \tag{5.60}$$

where E and E' are the energies of the incoming and outgoing leptons in the laboratory frame, $v = E - E'$. From relativistic kinematics, the final-state lepton energy for a given Q^2 and W can be expressed as

$$E' = E - \frac{W^2 - M^2 + Q^2}{2M}. \tag{5.61}$$

The helicity-dependent cross sections are computed from the squared amplitudes for the relevant resonance production helicity states:

$$\sigma_{L,R}(q^2, W) = \frac{\pi W}{\kappa M} \cdot \frac{1}{2} \left(|A_{\pm 3}|^2 + |A_{\pm 1}|^2 \right), \tag{5.62}$$

$$\sigma_S(q^2, W) = \frac{\pi W}{\kappa M} \left(\frac{-Q^2}{q^2} \right) \frac{M^2}{W^2} \cdot \frac{1}{2} \left(|A_{0+}|^2 + |A_{0-}|^2 \right), \tag{5.63}$$

where the A_k denote the helicity amplitudes for the (n, π^+) final state. In the single-resonance approximation (e.g. for $\Delta(1232)$), we set $|A_k|^2 = |a_k(N^*)|^2$, where

$$a_k^{CC}(N^*) = f_k^{CC}(v + n \to N^*) \times \eta(N^* \to n + \pi^+), \tag{5.64}$$

and the η factor encodes the decay amplitude, including the resonance line shape (e.g. a Breit-Wigner distribution). To compute the charged-current production amplitudes f_k^{CC}, we define the following intermediate quantities:

$$R^V = \sqrt{2} \frac{M}{W} \frac{(W + M)Q}{(W + M)^2 - q^2} G^V(q^2),$$

$$R^A = \frac{Z\sqrt{2}}{6W} \left(W + M + \frac{2n\Omega W}{(W + M)^2 - q^2} \right) G^A(q^2),$$

$$C = \frac{Z}{6MQ} \left(W^2 - M^2 + n\Omega \frac{W^2 - M^2 + q^2}{(W + M)^2 - q^2} \right) G^A(q^2),$$

$$R^{\pm} = -(R^V \pm R^A),$$

where Z is a normalization constant for the axial current (taken as $Z = 0.76$ in GENIE [1]), and Ω is a model parameter set to $\Omega = 1.05\,\text{GeV}^2$ [17]. The form factors $G^V(q^2)$ and $G^A(q^2)$ are given by

$$G^{V,A}(q^2) = \left(1 - \frac{q^2}{4M^2}\right)^{1/2} \left(\frac{1}{1 - q^2/M_{V,A}^2}\right)^2, \tag{5.65}$$

with $M_V = 0.84\,\text{GeV}$ and $M_A = 1.12\,\text{GeV}$ representing the vector and axial masses, respectively, extracted from experiments.

For the specific case of the $P_{33}(1232)$ (or $\Delta(1232)$) resonance produced via charged-current interaction on a neutron, the Feynman-Kislinger-Ravndal (FKR) quark model gives the following helicity amplitudes:

$$f_{-3} = \sqrt{6}\,R^-, \tag{5.66}$$

$$f_{-1} = \sqrt{2}\,R^-, \tag{5.67}$$

$$f_{+1} = -\sqrt{2}\,R^+, \tag{5.68}$$

$$f_{+3} = -\sqrt{6}\,R^+, \tag{5.69}$$

$$f_{0+} = -2\sqrt{2}\,C, \tag{5.70}$$

$$f_{0-} = -2\sqrt{2}\,C. \tag{5.71}$$

To compute the cross sections, these amplitudes must be squared and substituted into the expressions for $\sigma_{L,R,S}$.

Now, we only need the Breit-Wigner decay amplitude squared for the resonance (mass $M^r = 1232$ MeV, width $\Gamma = 124$ MeV), which factorizes completely in the FKR model for a single resonance:

$$\eta^2(W) = \frac{\Gamma}{2\pi} \frac{1}{(W - M)^2 + \Gamma^2/4} \cdot \frac{1}{N}. \tag{5.72}$$

Note that a nuclear resonance may have multiple decay modes, and therefore, the appropriate branching ratio must be taken into account. The constant N is a normalization factor, estimated numerically such that

$$N = \int_{W_{\min}}^{W_{\max}} \eta^2(W)\,dW. \tag{5.73}$$

The implementation of this model in code is rather lengthy but again straightforward. While the production amplitude is only detailed in this chapter for the resonance $P_{33}(1232)$, the original papers by Rein and Sehgal provide expressions for a variety of other resonances.

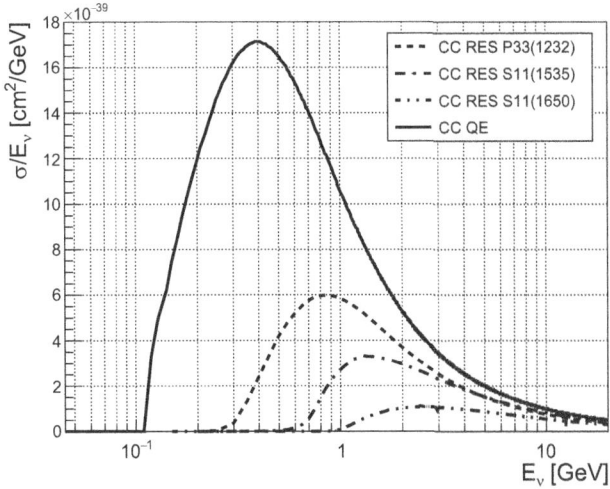

Fig. 5.9 The Llewellyn Smith charge-current (CC) quasi-elastic (QE) cross section divided by the neutrino energy for the process $\nu_\mu + n \to p + \mu^-$ is shown together with a few of the Rein-Sehgal CC resonance (RES) single-pion resonance cross sections. Examples shown for the resonances P33(1232), S11(1535), and S11(1650)

To obtain the total cross section as a function of the neutrino energy, a two-dimensional numerical integration is required, since the expression is doubly differential in Q^2 and W. When scanning with the neutrino energy, we can calculate the appropriate Q^2 and W limits based on the expressions we derived earlier. The result of this integration, showing the charged-current (CC) single-pion resonant (RES) cross section from a few single resonances, is plotted in Fig. 5.9 together with the earlier CC quasi-elastic (QE) cross-section result for neutrino scattering. The single-pion resonance cross-section values clearly become comparable to that of CCQE already at $E_\nu \approx 1$ GeV.

5.4.2 Final-State Kinematics

To implement an event generator based on the Rein-Sehgal cross section, the hadronic invariant mass W and the (negative) squared momentum transfer Q^2 must first be randomly sampled within their physical limits, since the cross section is differential in these variables. These physical boundaries were derived earlier in Sect. 5.2.1. However, unlike previous examples where phase-space generation was automatically handled by TGenPhaseSpace, a better approach is to sample W and Q^2 directly, making sure that the final-state kinematics is uniformly sampled.

Once W and Q^2 are sampled within their physical bounds uniformly, the kinematics of the final-state lepton can be constructed in the rest frame of the target nucleon. From Eq. 5.13, the lepton energy E_l is related to the incoming neutrino energy E_ν, the nucleon mass, and the kinematic variables W and Q^2. It is convenient to define the components of the lepton momentum parallel and perpendicular to the incoming neutrino direction. If we adopt a coordinate system in which the incoming neutrino four-momentum is $p_\nu = (E_\nu, 0, 0, E_\nu)$, the final-state lepton momentum is $p_l = (E_l, p_x, p_y, p_\parallel)$, with $p_x^2 + p_y^2 = |\mathbf{p}_\perp|^2$ and $p_\parallel$ aligned with the beam axis. The following kinematic relations apply

$$q^2 = (p_\nu - p_l)^2 = m_l^2 - 2p_\nu p_l$$

$$Q^2 = -q^2 = 2E_\nu(E_l - |\mathbf{p}_\parallel|) - m_l^2$$

$$|\mathbf{p}_\parallel| = E_l - \frac{Q^2 + m_l^2}{2E_\nu}.$$

With E_l and $|\mathbf{p}_\parallel|$ known, the transverse component $|\mathbf{p}_\perp|$ follows from

$$|\mathbf{p}_\perp| = \sqrt{E_l^2 - |\mathbf{p}_\parallel|^2 - m_l^2}.$$

A random azimuthal angle ϕ is then selected to compute p_x and p_y.

To generate the nucleon and pion final-state momenta, we use the known four-momenta of the incoming neutrino, target nucleon, and final-state lepton. The four-momentum of the intermediate resonance in the reaction $\nu + n \rightarrow N^* + \mu$ is given by

$$p_{N^*} = p_n + p_\nu - p_\mu. \tag{5.74}$$

Finally, we can use the `TGenPhaseSpace` class to simulate the decay of the resonance—assumed to have four-momentum p_{N^*} and mass $m_{\Delta(1232)}$—into a nucleon and a pion, as described in previous chapters. Since the implementation mirrors earlier examples, we do not repeat the code here.

The output of such an event generator is shown in Fig. 5.10, including the reconstructed hadronic invariant mass W, calculated via $W^2 = (p_\pi + p_p)^2$, and the final-state pion momentum distribution. As expected, the W distribution peaks at the $\Delta(1232)$ mass, and the pions exhibit a momentum distribution consistent with the estimates derived earlier from kinematics.

To reconstruct the neutrino energy for final states involving a single pion, we use a modified version of Eq. 5.43, accounting for the additional pion:

$$E_\nu = \frac{M_p^2 - (M_n - E_b - E_l - E_\pi)^2 + |\mathbf{p}_l + \mathbf{p}_\pi|^2}{2\left[(M_n - E_b - E_l - E_\pi) + |\mathbf{p}_l + \mathbf{p}_\pi|\cos\theta_\nu\right]}. \tag{5.75}$$

This concludes our chapter on the theoretical foundations and practical implementation of some elements of neutrino-nucleon scattering kinematics. The topic is extensive, and we have deliberately omitted several complex effects—including final-state interactions (FSI) of pions, two-particle-two-hole (2p2h) processes,

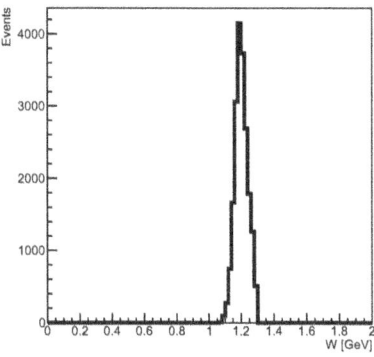 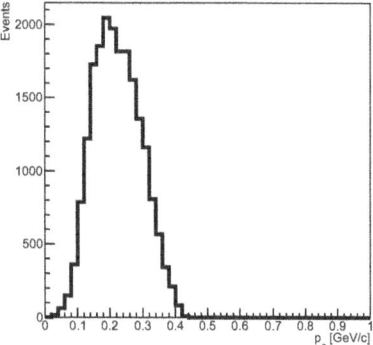

Fig. 5.10 Final-state kinematics from the CC single-pion resonance (RES) event generator for the process $\nu_\mu + n \rightarrow \Delta(1232) \rightarrow p + \mu + \pi$, using an initial neutrino energy of $E_\nu = 0.6\,\text{GeV}$ and a stationary target nucleon

Coulomb corrections, random phase approximation (RPA), and more sophisticated models of the nuclear initial state—to keep the discussion and accompanying code focused and manageable. Many of these effects are implemented in modern neutrino event generators and can be studied in greater detail with those tools.

In the final chapter, we turn to the intriguing phenomenon of neutrino flavour oscillations. We will apply our charged-current scattering codes to simulate neutrino scattering events in the far detector of the Tokai-to-Kamioka (T2K) experiment, thereby illustrating how neutrino flavour oscillations manifest in the phenomenon of electron neutrino appearance in a muon neutrino beam.

References

1. C. Andreopoulos et al., The GENIE neutrino Monte Carlo generator. Nuclear Instrum. Methods **A614**, 87–104 (2010)
2. Y. Hayato, *Neut* Nuclear Phys. B: Proc. Suppl. (112)(1–3), 171–176 (2002)
3. T. Golan, J.T. Sobczyk, J. Zmuda NuWro: the Wrocław Monte Carlo generator of neutrino interactions. Nucl. Phys. Proc. Suppl. **499**, 229–32 (2012)
4. O. Buss et al., Transport-theoretical description of nuclear reactions. Phys. Rep. **512**, 1–124 (2012)
5. U. Mosel, Neutrino interactions with nucleons and nuclei: importance for long-baseline experiments. Annu. Rev. Nucl. Part. Sci. **66**, 171–95 (2016)
6. J.A. Formaggio, G.P. Zeller, From eV to EeV: Neutrino cross sections across energy scales. Rev. Mod. Phys. **84**, 1307 (2012)
7. C.L. Cowan, F. Reines et al., Detection of the free neutrino. Science **124**, 3212 (1956)
8. C.H.L. Smith, Neutrino reactions at accelerator energies. Phys. Rept. **3C**, 261 (1972)
9. H. Budd, A. Bodek, J. Arrington, Modeling quasi-elastic form factors for electron and neutrino scattering (2003). arXiv:hep-ex/0308005
10. V. Bernard, L. Elouadrhiri, Meißner, Ulf-G. Axial structure of the nucleon. J. Phys. G: Nucl. Part. Phys. **28**, R1 (2002)
11. R.D. Woods, D.S. Saxon, Diffuse surface optical model for nucleon-nuclei scattering. Phys. Rev. **95**, 577 (1954)

12. A. Bodek, J.L. Ritchie, Fermi-motion effects in deep-inelastic lepton scattering from nuclear targets. Phys. Rev. D **23**, 1070 (1981)
13. O. Benhar, A. Fabrocini, S. Fantoni, I. Sick, Spectral function of finite nuclei and scattering of GeV electrons. Nucl. Phys. A **579**, 493–517 (1994)
14. R.A. Smith, E.J. Moniz, Neutrino reactions on nuclear targets. Nucl. Phys. **B43**, 605 (1972)
15. R. Gran, J. Nieves, F. Sanchez, M.J.V. Vacas, Neutrino-nucleus quasi-elastic and 2p2h interactions up to 10 GeV. Phys. Rev. D **88**, 113007 (2013)
16. A. Bodek, T. Cai, Removal energies and final state interaction in lepton nucleus scattering. Eur. Phys. J. C **79**, 293 (2019)
17. D. Rein, L.M. Sehgal, Neutrino-excitation of Baryon resonances and single pion production. Ann. Phys. **133**, 79–153 (1981)
18. R.P. Feynman, M. Kislinger F. Ravndal, Current matrix elements from a relativistic quark model. Phys. Rev. D **3**, 2706 (1971)
19. F. Ravndal, Weak production of nuclear resonances in a relativistic quark model. Nuovo Cimento A **18**, 385 (1973)

Neutrino Flavour Oscillations

6

Abstract

In this chapter, we explore the phenomenon of neutrino flavour oscillations, with particular emphasis on long-baseline experiments. We numerically solve the system of coupled differential equations governing neutrino propagation in both vacuum and matter and use these solutions to compute the flavour transition probabilities and various matter effects. Building on the event generation techniques developed in previous chapters, we then apply the rejection sampling algorithm to simulate charged-current quasi-elastic (CCQE) neutrino scattering events using a far detector neutrino beam flux model, in order to reproduce the appearance spectrum of electron neutrinos emerging from a muon-neutrino-dominated beam.

6.1 Neutrino Flavour Oscillation

Although the underlying origin of neutrino flavour transitions during propagation remains unknown, an analogous phenomenon is well understood in the quark sector, where flavour mixing arises from Yukawa interactions[1] with the Higgs condensate[2] [1]. If neutrinos possess mass and lepton flavours mix similarly, then lepton flavour conservation is violated during propagation—a phenomenon known as *neutrino flavour oscillation*.[3]

In the standard model, charged leptons acquire mass via Yukawa couplings with the Higgs field, H:

---

[1] Named after Japanese theoretical physicist H. Yukawa (1907–1981).

[2] Named after British theoretical physicist P. W. Higgs (1929–2024).

[3] A nice review of the history of neutrino oscillations is given in the article S. M. Bilenky *Bruno Pontecorvo and Neutrino Oscillations* Advances in High Energy Physics 2013, 873236 (2013).

© The Author(s), under exclusive license to Springer Nature Switzerland AG 2026   147
B. Radics, *Neutrino Physics*, Lecture Notes in Physics 1043,
https://doi.org/10.1007/978-3-032-03993-4_6

$$\mathcal{L}_\ell = -Y^{ij} \bar{L}_L^i H e_R^j + \text{h.c.}, \tag{6.1}$$

where Y^{ij} is the Yukawa matrix containing the lepton-Higgs couplings, $L_L^i = (\nu_L^i, e_L^i)^T$ the left-handed lepton doublet for generation i, and e_R^j the right-handed lepton singlet for generation j. When the Higgs field acquires a vacuum expectation value $\langle H \rangle = (0, \text{v}/\sqrt{2})^T$, the mass term becomes

$$\mathcal{L}_\ell = -\frac{\text{v}}{\sqrt{2}} Y^{ij} \bar{e}_L^i e_R^j + \text{h.c.} \tag{6.2}$$

with the charged lepton mass matrix $M_\ell = \frac{\text{v}}{\sqrt{2}} Y$, a generally non-diagonal matrix. The charged lepton mass matrix M_ℓ can be diagonalized via unitary matrices V^ℓ and V_R^ℓ as

$$V^{\ell\dagger} M_\ell V_R^\ell = M_\ell^{\text{diag}} = \text{diag}(m_e, m_\mu, m_\tau), \tag{6.3}$$

such that V^ℓ rotates the left-handed fields: $e_L \to V^{\ell\dagger} e_L$.

If neutrinos possess mass, and they are of Dirac nature, a similar procedure yields

$$\bar{\nu}_L M_\nu \nu_R = \bar{\nu}_L V^\nu V^{\nu\dagger} M_\nu V_R^\nu V_R^{\nu\dagger} \nu_R = \bar{\nu}_L' V^{\nu\dagger} M_\nu V_R^\nu \nu_R', \tag{6.4}$$

so that the neutrino fields are rotated to their mass basis, yielding the diagonal mass matrix $V^{\nu\dagger} M_\nu V_R^\nu = M_\nu^{\text{diag}} = \text{diag}(m_1, m_2, m_3)$. This implies the definition $\nu_L' = V^{\nu\dagger} \nu_L$ and $e_L' = V^{\ell\dagger} e_L$ for the transformations of the fields from the interaction basis to the mass basis.

Experimentally, neutrino flavour is identified through charged-current interactions. The corresponding interaction Lagrangian in the interaction basis is given by

$$\mathcal{L}_{CC} = -\frac{g}{\sqrt{2}} \bar{e}_L \gamma^\mu W_\mu \nu_L + \text{h.c.}. \tag{6.5}$$

After rotating the fields into their mass basis, the Lagrangian becomes

$$\mathcal{L}_{CC} = -\frac{g}{\sqrt{2}} \bar{e}_L' \gamma^\mu W_\mu \underbrace{(V^{\ell\dagger} V^\nu)}_{U} \nu_L' + \text{h.c.}, \tag{6.6}$$

where the lepton mixing matrix U can be identified as

$$U = V^{\ell\dagger} V^\nu. \tag{6.7}$$

Here, U is a 3×3 unitary matrix assuming three active, light neutrinos.

The elements of the U matrix are typically parametrized (see next sections) and extracted from experimental data. Understanding the structure of U remains a major goal in particle physics, with important implications for unified models of lepton and quark mixing [2].

6.2 Neutrino Propagation

6.2.1 Neutrino Propagation in Vacuum

Neutrino flavour is identified experimentally through the flavour of the associated charged lepton produced in charged-current interactions. The phenomenon of neutrino oscillations implies that the weak interaction eigenstates of neutrinos, $|\nu_\alpha\rangle$ with $\alpha = e, \mu, \tau$, are linear combinations of mass eigenstates, $|\nu_i\rangle$ with $i = 1, 2, 3$. Assuming three active neutrino species, a neutrino flavour state as a function of time is expressed as

$$|\nu_\alpha(t)\rangle = \sum_{i=1}^{3} U_{\alpha i}^* |\nu_i(t)\rangle, \tag{6.8}$$

where U is a unitary matrix.[4]

The probability of detecting a neutrino in flavour state β, given that it was produced in flavour state α and has propagated for a time t in vacuum, is given by projecting the time-evolved state onto $|\nu_\beta\rangle$:

$$P_{\nu_\alpha \to \nu_\beta}(t) = |\langle \nu_\beta | |\nu_\alpha(t)\rangle|^2 = |\sum_{i=1}^{3}\sum_{j=1}^{3} U_{\alpha i}^* U_{\beta j} \langle \nu_j | \nu_i(t)\rangle|^2. \tag{6.9}$$

Assuming plane-wave propagation of the mass eigenstates, $|\nu_i(t)\rangle = e^{-iE_i t}|\nu_i\rangle$, this becomes

$$P_{\nu_\alpha \to \nu_\beta}(t) = \sum_{i,j=1}^{3} U_{\alpha i} U_{\beta i}^* U_{\alpha j}^* U_{\beta j} e^{-i(E_i - E_j)t}. \tag{6.10}$$

Defining the shorthand $J_{\alpha\beta}^{ij} \equiv U_{\alpha i} U_{\beta i}^* U_{\alpha j}^* U_{\beta j}$, the oscillation probability for antineutrinos becomes

$$P_{\bar{\nu}_\alpha \to \bar{\nu}_\beta}(t) = \sum_{i,j=1}^{3} J_{\alpha\beta}^{*ij} e^{-i(E_i - E_j)t}, \tag{6.11}$$

---

[4] The definition of the field rotation direction is arbitrary and does not affect the physical observables as long as consistency is maintained.

since a CP transformation involves complex conjugation of the spinor fields and hence of the U matrix.[5] The difference between neutrino and antineutrino oscillation probabilities due to CP violation arises from the imaginary part of the product of matrix elements in $J_{\alpha\beta}^{ij}$. A parametrization-independent and phase-convention-independent measure of CP violation is given by the *Jarlskog invariant*[6] [3], defined as

$$J_{CP} = \text{Im}[U_{\alpha i} U_{\beta j} U_{\alpha j}^* U_{\beta i}^*], \tag{6.12}$$

for any distinct flavour indices $\alpha \neq \beta$ and mass indices $i \neq j$. Even though it is independent of the parametrization, it is often expressed using a standard parametrization, see Sect. 6.2.3. A nonzero value of J_{CP} indicates genuine CP violation in neutrino oscillations (or, equivalently, in the quark sector) arising from the mixing matrices. It is evident that if U is real ($U = U^*$), then J_{CP} is zero—there is no CP violation in neutrino oscillations in that case. However, $J_{CP} \neq 0$ is not enough to get observable CP violation effects.

Assuming neutrinos are ultrarelativistic, we can approximate their energies as $E_i \simeq E + \frac{m_i^2}{2E}$, so that

$$E_i - E_j \simeq \frac{\Delta m_{ij}^2}{2E}, \quad \text{where} \quad \Delta m_{ij}^2 = m_i^2 - m_j^2. \tag{6.13}$$

Using $t \simeq L$ in natural units ($c = 1$), the phase of the complex exponential can be expressed as

$$(E_i - E_j)t \simeq \left(\frac{\Delta m_{ij}^2 t}{2E} \right) = 2.534 \frac{\Delta m_{ij}^2}{[\text{eV}^2]} \frac{L/E}{[\text{km/GeV}]}. \tag{6.14}$$

This expression implies that if neutrino flavour oscillations are observed, at least some neutrino masses must be nonzero. Therefore, observable CP violation requires both a nonzero Jarlskog invariant and nonzero mass-squared differences. Notice that neutrino oscillation experiments are sensitive only to mass-squared differences, not to the absolute mass scale.

6.2.2 Neutrino Propagation in Matter

The propagation of neutrinos in matter, assuming three-flavour states and corresponding masses m_1, m_2, and m_3, is described in the flavour basis by the following

---

[5] A charge-conjugated Dirac spinor is $\nu^c = i\gamma^2 \nu^*$, and parity transformation is given by γ^0, so CP transformation yields $\bar{\nu} = i\gamma^0\gamma^2\nu^*$.

[6] Named after Swedish theoretical physicist C. Jarlskog.

complex system of ordinary differential equations (ODEs) [4, 5]:

$$
\hbar i \frac{d\Psi}{dt} = \frac{1}{2E} \left[U \begin{pmatrix} m_1^2 & 0 & 0 \\ 0 & m_2^2 & 0 \\ 0 & 0 & m_3^2 \end{pmatrix} U^\dagger + 2\sqrt{2} G_F N(t) E \begin{pmatrix} 1 & 0 & 0 \\ 0 & 0 & 0 \\ 0 & 0 & 0 \end{pmatrix} \right] \Psi. \tag{6.15}
$$

Here, $\Psi = (\psi_1, \psi_2, \psi_3)$ is the neutrino flavour state vector, t denotes time, $\hbar$ is the reduced Planck constant,[7] E is the neutrino energy, and U is the 3×3 neutrino mixing matrix (whose parametrization will be discussed in the next subsection). G_F is the Fermi coupling constant, and $N(t)$ represents the (possibly varying) number density of electrons in the medium traversed by the neutrino.

The right-hand side of the equation contains two contributions. The first term describes flavour oscillations in vacuum, with mixing driven by the off-diagonal elements of the matrix U. If U were the identity matrix, there would be no flavour mixing. Moreover, if the neutrino masses were all zero, or if their squared mass differences vanished (i.e. fully degenerate masses), there would be no observable flavour transitions during propagation.

The second term accounts for coherent forward scattering via the weak charged-current interaction in matter, resulting in an effective potential $V_W = \sqrt{2} G_F N$. This matter potential modifies the Hamiltonian eigenvalues and thus alters both the effective masses and the mixing angles relative to vacuum, as we shall confirm later.[8] Depending on the matter density along the propagation path—such as within the Earth or the Sun—these effects can be significant.

To estimate the order of magnitude of the weak potential, V_W, relative to the vacuum oscillation term $\Delta m^2/4E$, we begin by estimating the electron density in rock. Consider a typical rock such as quartz (SiO_2), a common silicate mineral, with an average mass density of $\rho_E \approx 2.7 \, \text{g/cm}^3$. The atomic masses of silicon and oxygen are approximately 28 and 16 g/mol, respectively, yielding a molar mass for SiO_2 of $M = 28 + 2 \cdot 16 = 60 \, \text{g/mol}$. The number density of SiO_2 molecules is then given by

$$
N_{SiO_2} = \frac{\rho}{M} \times N_A = \frac{2.7 \, \text{g/cm}^3}{60 \, \text{g/mol}} \times 6.022 \times 10^{23} \, \text{mol}^{-1} \approx 2.7 \times 10^{22} \, \text{cm}^{-3}. \tag{6.16}
$$

To obtain the electron density, we count the number of electrons per SiO_2 molecule. With atomic numbers 14 and 8 for Si and O, respectively, the total number of electrons per molecule is $14 + 2 \cdot 8 = 30$. Thus, the estimated electron number density is

---

[7] Named after German theoretical physicist M. Planck (1858–1947).

[8] All neutrino flavours also experience a neutral-current potential $V_Z = \sqrt{2} G_F N$ due to scattering on electrons and quarks, but since it contributes equally to all flavours, it does not affect the oscillation probabilities in long-baseline experiments.

$$N_e = 2.7 \times 10^{22}\,\text{cm}^{-3} \times 30 \approx 8.1 \times 10^{23}\,\text{cm}^{-3}.$$

An alternative method yields a similar result. Noting that the average atomic mass per electron in typical rock is approximately $\langle A/Z \rangle \sim 2$, we find

$$N_e = \rho \times N_A \times \frac{1}{\langle A/Z \rangle} = 2.7\,\text{g/cm}^3 \times 6.022 \times 10^{23}\,\text{mol}^{-1} \times \frac{1}{2\,\text{g/mol}} \approx 8.13 \times 10^{23}\,\text{cm}^{-3}.$$

Using this estimate for the electron density in rock, along with $G_F \simeq 8.96 \times 10^{-38}\,\text{eV} \cdot \text{cm}^3$, $\Delta m^2 \simeq 10^{-4}\,\text{eV}^2$, and a neutrino energy of $E = 100\,\text{MeV}$, we find

$$V_W = \sqrt{2} G_F N_e \sim 1.0 \times 10^{-13}\,\text{eV}, \tag{6.17}$$

$$\frac{\Delta m^2}{4E} = \frac{10^{-4}\,\text{eV}^2}{4 \cdot 100\,\text{MeV}} = 2.5 \times 10^{-13}\,\text{eV}. \tag{6.18}$$

However, if the neutrino propagation distance is much shorter than the characteristic oscillation length, $L \ll L_{\text{osc}} = 4\pi E/|\Delta m^2|$, the matter effect may be negligible. For terrestrial neutrino experiments, several approximations are typically employed: Matter effects can often be ignored in short-baseline or reactor experiments; a constant matter density is a reasonable assumption for atmospheric or long-baseline accelerator experiments with baselines of a few hundred kilometres, while variable density must be taken into account for longer baselines of several thousand kilometres, such as those relevant for neutrino telescopes at the South Pole or very-long-baseline experiments. In what follows, we will consider the case of a weak potential arising from charged-current interactions of electron neutrinos with bound electrons in matter, which becomes relevant in long-baseline accelerator experiments.

It is convenient to rewrite the evolution equation in terms of the propagation distance rather than time. Assuming that the neutrino masses are negligible compared to their energies, we can approximate their propagation speed as the speed of light, c, and make the substitution $t \to L = ct$. The Schrödinger-like evolution equation then becomes

$$\frac{d\Psi}{dL} = -i\frac{1}{2E\hbar c}\left[U \begin{pmatrix} m_1^2 & 0 & 0 \\ 0 & m_2^2 & 0 \\ 0 & 0 & m_3^2 \end{pmatrix} U^\dagger + 2\sqrt{2}G_F N_e(L)E \begin{pmatrix} 1 & 0 & 0 \\ 0 & 0 & 0 \\ 0 & 0 & 0 \end{pmatrix} \right]\Psi. \tag{6.19}$$

A natural choice of basis for the state vector Ψ is the flavour basis, defined by

$$|\nu_e\rangle = \begin{pmatrix} 1 \\ 0 \\ 0 \end{pmatrix}, \quad |\nu_\mu\rangle = \begin{pmatrix} 0 \\ 1 \\ 0 \end{pmatrix}, \quad |\nu_\tau\rangle = \begin{pmatrix} 0 \\ 0 \\ 1 \end{pmatrix}.$$

In this basis, a general neutrino state is given by

$$|\Psi\rangle = c_1 \begin{pmatrix} 1 \\ 0 \\ 0 \end{pmatrix} + c_2 \begin{pmatrix} 0 \\ 1 \\ 0 \end{pmatrix} + c_3 \begin{pmatrix} 0 \\ 0 \\ 1 \end{pmatrix} \equiv \begin{pmatrix} \psi_1 \\ \psi_2 \\ \psi_3 \end{pmatrix}, \tag{6.20}$$

where $c_1, c_2, c_3 \in \mathbb{C}$ are complex coefficients normalized such that $|c_1|^2 + |c_2|^2 + |c_3|^2 = 1$.

6.2.3 Standard Parametrization of the Mixing Matrix

If there are three active neutrinos, assumed to be Dirac particles, the unitary mixing matrix U can be parametrized by three mixing angles and a single complex phase. This so-called *standard parametrization* [6], also referred to as the Pontecorvo-Maki-Nakagawa-Sakata[9] (PMNS) parametrization[7, 8], takes the form

$$U = \begin{pmatrix} 1 & 0 & 0 \\ 0 & c_{23} & s_{23} \\ 0 & -s_{23} & c_{23} \end{pmatrix} \begin{pmatrix} c_{13} & 0 & s_{13}e^{-i\delta_{CP}} \\ 0 & 1 & 0 \\ -s_{13}e^{i\delta_{CP}} & 0 & c_{13} \end{pmatrix} \begin{pmatrix} c_{12} & s_{12} & 0 \\ -s_{12} & c_{12} & 0 \\ 0 & 0 & 1 \end{pmatrix}, \tag{6.21}$$

where $c_{ij} = \cos\theta_{ij}$ and $s_{ij} = \sin\theta_{ij}$ for $i, j \in \{1, 2, 3\}$, and δ_{CP} is the CP-violating phase. If neutrinos were Majorana particles, the number of physical degrees of freedom in the mixing matrix would increase to six due to the presence of two additional Majorana phases. We restrict our discussion to the case of Dirac neutrinos.

The previously introduced Jarlskog invariant, which quantifies CP violation in neutrino oscillations, can be expressed in terms of the standard parametrization as

$$J_{CP} = \cos\theta_{13} \sin 2\theta_{12} \sin 2\theta_{13} \sin 2\theta_{23} \sin \delta_{CP}. \tag{6.22}$$

The invariant J_{CP} vanishes whenever any of the mixing angles is zero or maximal (i.e. $\theta_{ij} = 0$ or $\pi/2$), or if the CP-violating phase δ_{CP} is zero or π.

6.3 Numerically Solving the ODE of Neutrino Propagation

We now have all the ingredients necessary to solve Eq. 6.19 numerically. The conceptual approach to evolving flavour oscillations over space is essentially the same as the complex-domain coupled ODE problem discussed in Sect. 2.3.2. The right-hand side of the evolution equation must be implemented using the complex algebra capabilities of the GNU Scientific Library (GSL). In addition, the model

---

[9] Named after Italian physicist B. Pontecorvo (1913–1993), and Japanese physicists Z. Maki (1929–2005), M. Nakagawa, and S. Sakata (1911–1970).

Table 6.1 Summary of parameters used in the neutrino oscillation ODE example calculations. The parameter values are taken from the Particle Data Group publication [1]

Quantity	Value
Δm_{21}^2	7.37×10^{-5} eV^2
Δm_{32}^2	2.437×10^{-3} eV^2
θ_{12}	$33.41°$
θ_{23}	$49.1°$
θ_{13}	$8.54°$
δ_{CP}	$-90°$

parameters must be defined: the mixing angles θ_{12}, θ_{23}, θ_{13}, the phase δ_{CP}, and the mass-squared differences Δm_{21}^2 and Δm_{32}^2.

For the mass ordering, we follow the convention adopted by the Particle Data Group (PDG). In the case of the *normal hierarchy* (NH), where $m_1 \ll m_2 < m_3$, the mass-squared values are defined as

$$m_2^2 = \Delta m_{21}^2, \qquad m_3^2 = \Delta m_{21}^2 + \Delta m_{32}^2.$$

In the case of the *inverted hierarchy* (IH), where $m_3 \ll m_1 < m_2$, we define

$$m_2^2 = |\Delta m_{32}^2|, \qquad m_1^2 = |\Delta m_{21}^2 + \Delta m_{32}^2|,$$

noting that Δm_{32}^2 is negative in this convention.

For the experimental setup, we consider representative baseline distances of $L = 300$ km and 1200 km, and neutrino energies of $E = 0.6$ GeV and 2.5 GeV, which are characteristics of the T2K and DUNE experiments, respectively. We will begin by solving the evolution equation in vacuum, and subsequently include the matter term to account for matter effects.

The mixing parameters used are taken from the best-fit values provided by the Particle Data Booklet, which in turn are based on the NuFIT global analysis [9]. These parameters are summarized in Table 6.1.

6.3.1 Propagation in Vacuum

To simulate neutrino propagation in vacuum, we solve a differential equation with only the vacuum term on the right-hand side:

$$\frac{d\Psi}{dL} = -i\frac{1}{2E\hbar c}\left[U \begin{pmatrix} m_1^2 & 0 & 0 \\ 0 & m_2^2 & 0 \\ 0 & 0 & m_3^2 \end{pmatrix} U^\dagger \right] \Psi. \qquad (6.23)$$

In the implementation, see code listing 6.1, we begin by defining the global model parameters and relevant numerical constants. We then create a function responsible for initializing the elements of the mixing matrix U, setting the mass-squared values, and evaluating the matrix algebra needed to construct the effective Hamiltonian in the flavour basis.

Note that complex numbers are handled using the `gsl_complex` type provided by GSL. The initialization routine is encapsulated in a function `void initialize()`, which fills the matrix elements of U, sets the mass matrix, and performs the matrix and other operations required to evaluate the right-hand side of Eq. 6.23.

Code listing 6.1 Part 1: Calculating the quantities on the right-hand side of the vacuum-only ODE for neutrino propagation during initialization

```cpp
#include <TFile.h>
#include <TMath.h>
#include <TGraph.h>
#include <TCanvas.h>
#include <TStyle.h>
#include <TAxis.h>
#include <TLegend.h>
#include <TMultiGraph.h>

#include <stdio.h>
#include <math.h>
#include <gsl/gsl_errno.h>
#include <gsl/gsl_odeiv2.h>
#include <gsl/gsl_complex.h>
#include <gsl/gsl_complex_math.h>

#include "Constants.h"

// Mixing angles in radian
double t12 = 33.41*kPi/180.0;
double t13 = 8.54*kPi/180.0;
double t23 = 49.1*kPi/180.0;
double delta = -90.0*kPi/180.0;

double c13 = cos(t13);
double s13 = sin(t13);
double c12 = cos(t12);
double s12 = sin(t12);
double c23 = cos(t23);
double s23 = sin(t23);

// Differences in mass squared
double dm32=2.437e-03; // eV^2
double dm21=7.37e-05; // eV^2
```

```
// Neutrino energy
double Enu = 0.6; // GeV

// 3x3 complex U, Uconj, and mass matrices
gsl_complex U[3][3], UconjT[3][3], Mass[3][3], H1[3][3], H[3][3];

void initialize(){

  // exp(i*delta) and exp(-i*delta)
  gsl_complex idelta = gsl_complex_rect(0,delta);
  gsl_complex imdelta = gsl_complex_rect(0,-delta);
  gsl_complex exp_idelta = gsl_complex_exp(idelta);
  gsl_complex exp_imdelta = gsl_complex_exp(imdelta);

  // e1, e2, e3
  U[0][0] = gsl_complex_rect(c12*c13,0);
  U[0][1] = gsl_complex_rect(s12*c13,0);
  U[0][2] = gsl_complex_mul_real(exp_imdelta, s13);

  // mu1, mu2, mu3
  U[1][0] =
↪  gsl_complex_sub(gsl_complex_rect(-s12*c23,0),gsl_complex_mul_real
   (exp_idelta, c12*s23*s13));
  U[1][1] =
↪  gsl_complex_sub(gsl_complex_rect(c12*c23,0),gsl_complex_mul_real
   (exp_idelta, s12*s23*s13));
  U[1][2] = gsl_complex_rect(s23*c13,0);

  // tau1, tau2, tau3
  U[2][0] =
↪  gsl_complex_sub(gsl_complex_rect(s12*s23,0),gsl_complex_mul_real
   (exp_idelta, c12*c23*s13));
  U[2][1] = gsl_complex_sub(gsl_complex_rect(-c12*s23,
↪  0),gsl_complex_mul_real(exp_idelta, s12*c23*s13));
  U[2][2] = gsl_complex_rect(c23*c13,0);

  // Constants to get the correct dimensions
  double p = 1.0/(2.0*hbarc*1.0e+06*1.0e-18*Enu*1.0e+09);

  // calculate U conjugate transpose, and initialize H and M
  for(int i = 0; i < 3; i++){
    for(int j = 0; j < 3; j++){
      UconjT[i][j] = U[j][i];
```

```
        H[i][j] = gsl_complex_rect(0,0);
        H1[i][j] = gsl_complex_rect(0,0);
        Mass[i][j] = gsl_complex_rect(0,0);
    }
}

// Transpose UconjT
for(int i = 0; i < 3; i++)
    for(int j = 0; j < 3; j++)
        UconjT[i][j] = gsl_complex_conjugate(UconjT[i][j]);

// Mass squared differences
Mass[0][0] = gsl_complex_rect(0,0);
Mass[1][1] = gsl_complex_rect(dm21, 0);
Mass[2][2] = gsl_complex_rect(dm21+dm32,0);

// Calculate H1= Mass*UconjT
for(int i = 0; i < 3; i++){
    for(int j = 0; j < 3; j++){
        gsl_complex s = gsl_complex_rect(0, 0);
        for (int k = 0; k < 3; k++){
            s = gsl_complex_add(s,gsl_complex_mul(Mass[i][k], UconjT[k][j]));
        }
        H1[i][j] = s;
    }
}

// Calculate Hamiltonian H = U*H1
for(int i = 0; i < 3; i++){
    for(int j = 0; j < 3; j++){
        for (int k = 0; k < 3; k++)
            H[i][j] = gsl_complex_add(H[i][j],
↪ gsl_complex_mul(U[i][k],H1[k][j]));
        H[i][j] = gsl_complex_mul_real(H[i][j], p);
    }
}

}
...
```

The matrices U and related quantities are implemented as two-dimensional arrays of the `gsl_complex` type, with matrix multiplication written out explicitly for transparency. This part of the computation only needs to be performed once. The denominator is also evaluated using various powers of 10, depending on the units employed (MeV, fm, and km). The reader is encouraged to verify the dimensional consistency of the expressions.

Next, we need to instruct GSL on how to connect the right-hand side of the differential equation to the left-hand side—that is, how the system of equations and the corresponding variables are related in the ODE. This is implemented in a dedicated function, `complex_ode_func`; see code listing 6.2.

Code listing 6.2 Part 2: Evaluating and connecting the terms on the left-hand side and right-hand side of the vacuum-only ODE for neutrino propagation

```
...
// Define the complex ODE system
int complex_ode_func(double t, const double y[], double dydt[], void
↪  *params) {
  (void)(t); /* avoid unused parameter warning */
  gsl_complex psi_1 = gsl_complex_rect(y[0], y[1]);
  gsl_complex psi_2 = gsl_complex_rect(y[2], y[3]);
  gsl_complex psi_3 = gsl_complex_rect(y[4], y[5]);

  gsl_complex imag = gsl_complex_rect(0, -1);

  // The right-hand side of the diff. equation
  // using the GSL algebraic functions
  gsl_complex dpsi_1dt = gsl_complex_mul(H[0][0],psi_1);
  dpsi_1dt = gsl_complex_add(dpsi_1dt, gsl_complex_mul(H[0][1],psi_2));
  dpsi_1dt = gsl_complex_add(dpsi_1dt, gsl_complex_mul(H[0][2],psi_3));
  gsl_complex dpsi_2dt = gsl_complex_mul(H[1][0],psi_1);
  dpsi_2dt = gsl_complex_add(dpsi_2dt, gsl_complex_mul(H[1][1],psi_2));
  dpsi_2dt = gsl_complex_add(dpsi_2dt, gsl_complex_mul(H[1][2],psi_3));
  gsl_complex dpsi_3dt = gsl_complex_mul(H[2][0],psi_1);
  dpsi_3dt = gsl_complex_add(dpsi_3dt, gsl_complex_mul(H[2][1],psi_2));
  dpsi_3dt = gsl_complex_add(dpsi_3dt, gsl_complex_mul(H[2][2],psi_3));

  // Multiply by -i
  dpsi_1dt = gsl_complex_mul(dpsi_1dt,imag);
  dpsi_2dt = gsl_complex_mul(dpsi_2dt,imag);
  dpsi_3dt = gsl_complex_mul(dpsi_3dt,imag);

  // Obtain the 6 real and complex parts of the state variables
  dydt[0] = GSL_REAL(dpsi_1dt);
  dydt[1] = GSL_IMAG(dpsi_1dt);
  dydt[2] = GSL_REAL(dpsi_2dt);
  dydt[3] = GSL_IMAG(dpsi_2dt);
  dydt[4] = GSL_REAL(dpsi_3dt);
```

```
    dydt[5] = GSL_IMAG(dpsi_3dt);

    return GSL_SUCCESS;
}
...
```

All we do in the second part of the code is to perform the matrix-vector multiplication on the right-hand side of the ODE, between the fully evaluated matrix of quantities and the state vector Ψ, and store the resulting six-dimensional state vector (three real and three complex components) in the array `dydt[]`.

The remaining steps, shown in code listing 6.3, are to instantiate the GSL ODE driver, choose the integration algorithm, set the experimental baseline, and specify the initial value of the state vector Ψ for a given neutrino flavour. The ODE propagation is then performed within a loop, and the results are saved to an output file.

What we store are the probabilities of the system being in each of the states $|\Psi_1\rangle$, $|\Psi_2\rangle$, or $|\Psi_3\rangle$. For example, the probability of being in state $|\Psi_1\rangle$ (which corresponds to a ν_e state, due to the choice of basis) after propagation over a distance L is given by $|\langle\Psi_1|\Psi(L)\rangle|^2$, which is simply computed using the call `gsl_complex_abs2(psi_1)`. This works because we are projecting the propagated state vector (expressed in the chosen basis) onto one of the basis vectors. Only the component of the state vector aligned with that basis vector contributes; all other components vanish due to orthogonality.

Code listing 6.3 Part 3: Performing the evolution steps for neutrino propagation in the main function, and saving the output

```
...
int main(void){

    // Call the initialize function
    initialize();

    // Step size in km
    double h = 0.5;

    // Define the GSL ODE system
    gsl_odeiv2_system sys = {complex_ode_func, NULL, 6, NULL};
    // The driver needs the system, stepping algo,
    // the step size and the abs, rel tolerances
    gsl_odeiv2_driver *d = gsl_odeiv2_driver_alloc_y_new(&sys,
    ↪   gsl_odeiv2_step_rk8pd, h, 1e-06, 1e-06);

    // Baseline
```

```cpp
double L = 0.0, L1 = 1000; // km

// Initial value of the system:
// RE(psi_1), IM(psi_1), RE(psi_2), IM(psi_2, RE(psi_3), IM(psi_3)
double y[6] = {0.0, 0.0, 1.0, 0.0, 0.0, 0.0};

// Save the intermediate step values
std::vector<double> L_v, prob_psi1_v, prob_psi2_v, prob_psi3_v;
unsigned int nsteps = 0;

// Stepping through the evolution of the ODE
for (double Li = 0; Li <= L1; Li += h) {
  // Save the current propagation length
  L_v.push_back(L);

  // Apply the ODE driver step
  int status = gsl_odeiv2_driver_apply(d, &L, Li, y);

  // Check for errors
  if (status != GSL_SUCCESS) {
    printf("error, return value=%d\n", status);
    break;
  }

  // Retrieve the current wave function values
  gsl_complex psi_1 = gsl_complex_rect(y[0],y[1]);
  gsl_complex psi_2 = gsl_complex_rect(y[2],y[3]);
  gsl_complex psi_3 = gsl_complex_rect(y[4],y[5]);

  // Save the probabilities at the intermediate step values
  prob_psi1_v.push_back(gsl_complex_abs2(psi_1));
  prob_psi2_v.push_back(gsl_complex_abs2(psi_2));
  prob_psi3_v.push_back(gsl_complex_abs2(psi_3));

  nsteps++;
}

gsl_odeiv2_driver_free(d);

// Create a ROOT output file to draw the result in graphs
// onto a canvas object
TFile * ofile = new TFile("out_neut_vac.root", "RECREATE");
// The canvas
TCanvas *c1 = new TCanvas("c1", "c1",10,65,700,500);
```

```
c1->cd();
c1->SetGridx();
c1->SetGridy();
c1->SetFrameBorderMode(0);

// Create a multigraph
TMultiGraph *mg = new TMultiGraph();
// Individual graphs
TGraph * gr_psi1_t = new TGraph(nsteps, &L_v[0], &prob_psi1_v[0]);
TGraph * gr_psi2_t = new TGraph(nsteps, &L_v[0], &prob_psi2_v[0]);
TGraph * gr_psi3_t = new TGraph(nsteps, &L_v[0], &prob_psi3_v[0]);
gr_psi1_t->SetName("Pe");gr_psi1_t->SetLineColor(kRed);
gr_psi2_t->SetName("Pmu");gr_psi2_t->SetLineColor(kBlack);
gr_psi3_t->SetName("Ptau");gr_psi3_t->SetLineColor(kBlue);

// Add and draw the graphs
mg->Add(gr_psi1_t);
mg->Add(gr_psi2_t);
mg->Add(gr_psi3_t);
mg->Draw("AL");
mg->GetXaxis()->SetTitle("L[km]");
mg->GetYaxis()->SetTitle("Probability of state |#nu_{x}>");
mg->GetYaxis()->SetRangeUser(-0.01, 1.01);

// Draw a legend with the labels
TLegend *leg = new
    TLegend(0.6375358,0.7136842,0.8997135,0.9010526,NULL,"brNDC");
TLegendEntry
    *entry=leg->AddEntry("Pe","|<#Psi_{e}|#Psi(L)>|^{2}","lpf");
leg->AddEntry("Pmu","|<#Psi_{#mu}|#Psi(L)>|^{2}","lpf");
leg->AddEntry("Ptau","|<#Psi_{#tau}|#Psi(L)>|^{2}","lpf");
leg->Draw();

// Write the Canvas to the output ROOT file
c1->Write();
ofile->Close();

return 0;
}
```

Our first numerical result for vacuum-only oscillations is shown in Fig. 6.1. We observe characteristic maxima and minima in the probability of being in each of the ν_e, ν_μ, or ν_τ flavour states as a function of propagation distance. At every step, the total probability sums to 1, as required by unitarity.

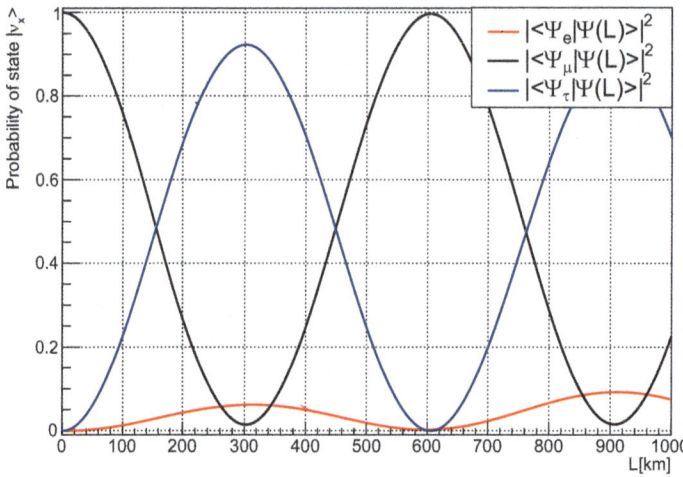

Fig. 6.1 Three-flavour neutrino oscillation probability in vacuum over a propagation length of 1000 km, using neutrino energy $E = 0.6\,\text{GeV}$, and an initial flavour state $|\Psi\rangle(L = 0) = (0, 1, 0) \equiv (0, \nu_\mu, 0)$

Historically, the Tokai-to-Kamioka (T2K) long-baseline neutrino experiment was designed around the oscillation maximum/minimum occurring at a propagation distance of approximately $L \approx 300$ km [10]. At this distance, we see an almost complete *disappearance* of the initial muon neutrino state, with conversion into tau neutrino flavour states at a level of roughly 93%, and into electron neutrino flavour states at around 6%. This phenomenon enabled the T2K experiment to achieve the first observation of electron neutrino *appearance* in an accelerator-generated muon neutrino beam [11], and the same baseline will be used by the upcoming Hyper-Kamiokande experiment [12].

We can change the neutrino beam energy to $E = 2.5$ GeV and compute the flavour oscillation pattern up to a propagation distance of $L = 2000$ km, keeping the initial neutrino flavour fixed to muon neutrino. The resulting behaviour is shown in Fig. 6.2. An oscillation maximum/minimum occurs at a distance of approximately $L \approx 1300$ km, which corresponds to the design baseline of the Deep Underground Neutrino Experiment (DUNE) [13]. This setup is intended to observe both the disappearance and appearance of different neutrino flavours in an accelerator-generated beam with a broad neutrino energy spectrum.

6.3.2 Oscillation in Neutrino Energy

Long-baseline neutrino experiments typically measure not only the flavour composition of a neutrino beam but also reconstruct the neutrino energy spectrum at various distances from the source. While the flavour composition evolves during propagation, it also depends on the neutrino energy.

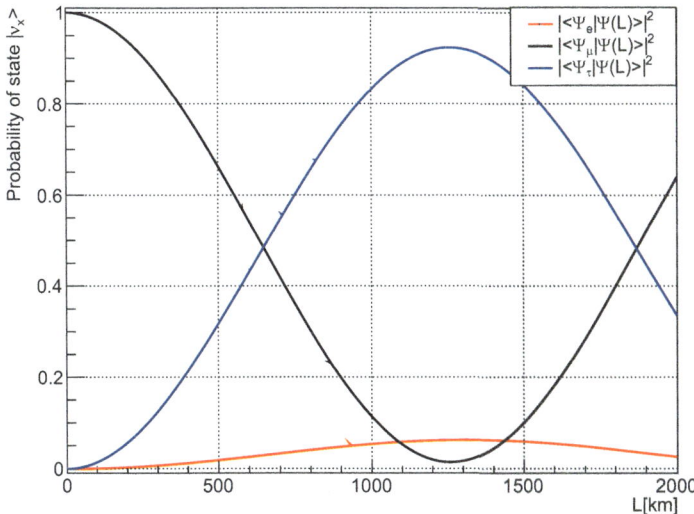

Fig. 6.2 Three-flavour neutrino oscillation probability in vacuum over a propagation length of 2000 km, using neutrino energy $E = 2.5\,\mathrm{GeV}$, and an initial flavour state $|\Psi\rangle(L = 0) = (0, 1, 0) \equiv (0, \nu_\mu, 0)$

With minimal modifications, we can reuse our vacuum neutrino propagation ODE code to study the energy dependence of the flavour oscillation probability. For the T2K experiment, we fix the baseline distance to $L = 300$ km and solve the ODE over a range of energies from $E = 0$ to 2 GeV. This can be implemented by wrapping the steps we took before in a loop over different energy values. The pseudocode, shown in code listing 6.4, illustrates how this can be achieved.

Code listing 6.4 Solving the vacuum ODE for neutrino propagation with a fixed baseline $L = 300$ km and varying energy

```
...
int main(void){

  // Step size in km
  double h = 0.5;
  // Abs, rel tolerances
  const double epsAbs = 1.0e-06;
  const double epsRel = 1.0e-06;

  // Energy range
  double Emin = 0.01, Emax = 2.0; // GeV
  int nsteps = 10000;
  double dE = (Emax-Emin)/(double)nsteps;
  // Baseline
```

```cpp
double L = 0.0; // km
double L1 = 300; // km


// Save the intermediate step values
std::vector<double> E_v, prob_psi1_v, prob_psi2_v, prob_psi3_v;

// Define the GSL ODE system
gsl_odeiv2_system sys = {complex_ode_func, NULL, 6, NULL};
gsl_odeiv2_driver *d = NULL;

// Scan of the energy
for (double Ei = Emin; Ei <= Emax; Ei += dE){
  // Set the neutrino energy
  Enu = Ei;
  printf("Neutrino E: %.5e\n", Enu);

  // Call the initialize function
  initialize();

  // Initial value of the system:
  // RE(psi_1), IM(psi_1), RE(psi_2), IM(psi_2), RE(psi_3), IM(psi_3)
  double y[6] = {0.0, 0.0, 1.0, 0.0, 0.0, 0.0};

  // The driver needs the system, stepping algo,
  // the step size and the abs, rel tolerances
  d= gsl_odeiv2_driver_alloc_y_new(&sys, gsl_odeiv2_step_rk8pd, h,
↪ epsAbs, epsRel);
  // reset initial distance to 0
  L = 0;

    // Apply the ODE driver step
    int status = gsl_odeiv2_driver_apply(d, &L, L1, y);

    // Check for errors
    if (status != GSL_SUCCESS) {
      printf("error, return value=%d\n", status);
      break;
    }

  // Retrieve the current wave function values
  gsl_complex psi_1 = gsl_complex_rect(y[0],y[1]);
  gsl_complex psi_2 = gsl_complex_rect(y[2],y[3]);
  gsl_complex psi_3 = gsl_complex_rect(y[4],y[5]);
```

```
// Save the probabilities at the intermediate step values
prob_psi1_v.push_back(gsl_complex_abs2(psi_1));
prob_psi2_v.push_back(gsl_complex_abs2(psi_2));
prob_psi3_v.push_back(gsl_complex_abs2(psi_3));

E_v.push_back(Enu);
}

gsl_odeiv2_driver_free(d);
...
```

The result of this procedure is shown in Fig. 6.3. The flavour oscillation probability exhibits sharp variations as a function of energy. Recall from Sect. 4.4, *Pion and kaon decay, accelerator neutrinos*, that the energy and spread of the neutrino beam are related to the angle relative to the direction of the decaying parent mesons. Consequently, the T2K experiment employs an off-axis configuration to produce a narrow-energy-band neutrino beam centred around $E = 0.6$ GeV. As is evident from Fig. 6.3, this energy corresponds to a peak in the appearance/disappearance probability of neutrino flavours.

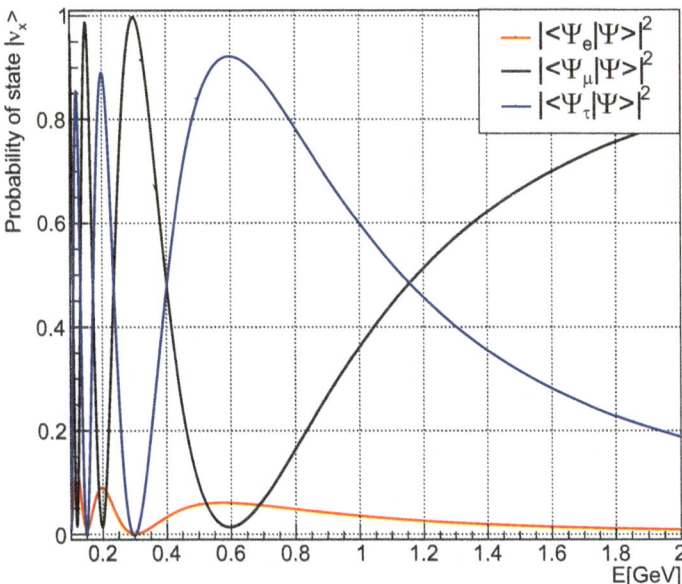

Fig. 6.3 Energy dependence of the three-flavour neutrino oscillation probability in vacuum with a fixed propagation length of 300 km, starting from an initial flavour state $|\Psi\rangle(L = 0) = (0, 1, 0) \equiv (0, \nu_\mu, 0)$

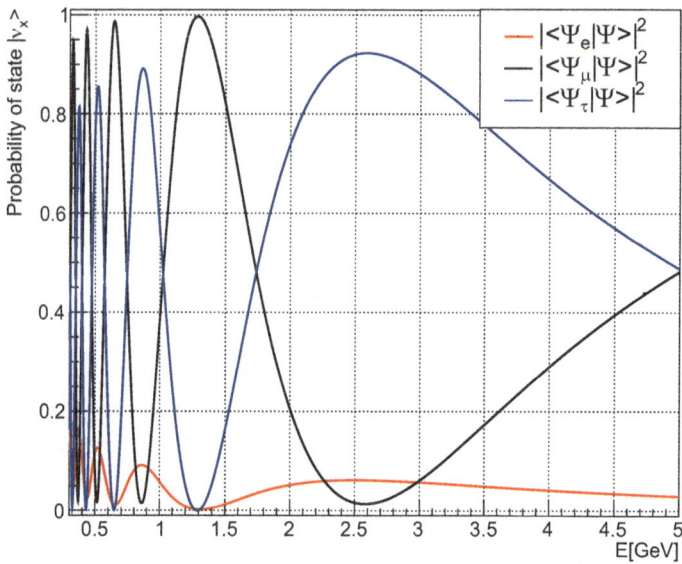

Fig. 6.4 Energy dependence of the three-flavour neutrino oscillation probability in vacuum with a fixed propagation length of 1300 km, starting from an initial flavour state $|\Psi\rangle(L = 0) = (0, 1, 0) \equiv (0, \nu_\mu, 0)$

For the DUNE baseline, the corresponding result is shown in Fig. 6.4, where the baseline distance is fixed at $L = 1300$ km and the neutrino energy ranges from $E = 0$ to 5 GeV. In this case, multiple maxima and minima in the oscillation probability appear around and above 1 GeV. DUNE is designed to use a wideband, on-axis neutrino beam to probe neutrino mixing parameters and other properties by leveraging the rich structure in the energy-dependent oscillation pattern.

Two important open questions in neutrino physics are: What type of mass hierarchy do neutrinos follow (normal or inverted hierarchy)? And what is the size of CP violation present in neutrino oscillations? Both effects can be explored with our code to study their impact on the oscillation pattern. In this section, we illustrate only the effect of CP violation, leaving the investigation of mass hierarchy as an exercise for the curious reader.

The impact of CP violation can be studied by comparing the oscillation probabilities for neutrinos and antineutrinos—specifically, by evaluating whether $P(\nu_\mu \to \nu_e) \stackrel{?}{=} P(\bar{\nu}_\mu \to \bar{\nu}_e)$—for different values of the CP-violating phase δ_{CP}, including both zero and nonzero cases. As discussed earlier, the antineutrino case requires taking the complex conjugate of the terms involving the U mixing matrix in Eq. 6.11.

The effect of varying the CP phase δ_{CP} is shown in Fig. 6.5 for electron neutrino appearance, assuming a $L = 300$ km propagation distance and an initial state of either $|\nu_\mu\rangle$ or $|\bar{\nu}_\mu\rangle$, with $\delta_{CP} = 0$ and $\delta_{CP} = -\pi/2$ radians. While the absolute probability for flavour change to $|\nu_e\rangle$ or $|\bar{\nu}_e\rangle$ remains relatively small, the relative

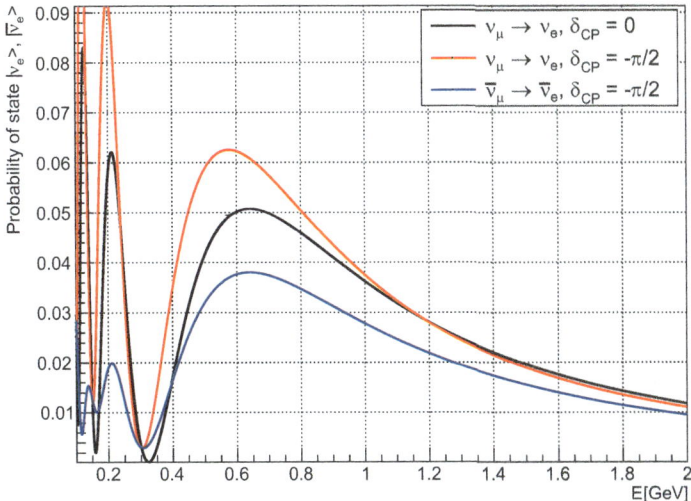

Fig. 6.5 Energy dependence of the three-flavour neutrino oscillation probability in vacuum with a fixed propagation length of 300 km, for electron (anti) neutrino appearance: $P(\nu_\mu \rightarrow \nu_e)$ and $P(\bar{\nu}_\mu \rightarrow \bar{\nu}_e)$. The cases of the CP-violating phase values $\delta_{CP} = 0$ and $-\pi/2$ are shown

change between the neutrino and antineutrino channels can reach nearly a factor of two when $\delta_{CP} = -\pi/2$, the value corresponding to maximal CP violation.

Since the true value of δ_{CP} is currently unknown, experiments scan over all possible values to determine which best fits the data. A recent result from the T2K experiment suggests a potentially large value of δ_{CP} [14]. Next-generation long-baseline neutrino experiments aim to perform precision measurements of the CP-violating phase and definitively determine the mass hierarchy.

6.3.3 Propagation in Matter with Constant Density

Over hundreds of kilometres of propagation, matter effects can significantly influence neutrino oscillation probabilities. These effects are included not only in long-baseline accelerator neutrino experiments but were originally motivated by solar neutrino observations. In the context of accelerator-based experiments on Earth, neutrinos propagate through the Earth's crust, which can be reasonably approximated as having constant density. This allows us to extend our code in a straightforward way by including the second term on the right-hand side of the ODE for neutrino propagation.

For constant matter density, we can treat N_e as independent of L, i.e. $N_e(L) \equiv N_e$, and write the ODE as

$$\frac{d\Psi}{dL} = -i\frac{1}{2E\hbar c}\left[U \begin{pmatrix} m_1^2 & 0 & 0 \\ 0 & m_2^2 & 0 \\ 0 & 0 & m_3^2 \end{pmatrix} U^\dagger + 2\sqrt{2}G_F N_e E \begin{pmatrix} 1 & 0 & 0 \\ 0 & 0 & 0 \\ 0 & 0 & 0 \end{pmatrix} \right] \Psi. \quad (6.24)$$

From this expression, we see that the energy dependence and the factor of 2 in the second term cancel, leaving only a constant additive term $\sqrt{2}G_F N_e$ that modifies the $(1, 1)$ element of the vacuum Hamiltonian matrix.

In terms of code implementation, the only additional step is to add a line at the end of the `void initialize()` function, where a new variable `gsl_complex MatPot` is defined for the matter potential. Its real part is set to the appropriate value discussed above and added as a complex contribution to the matrix element `H[0][0]`, as formally implemented in code listing 6.5.

Code listing 6.5 Including the matter potential in the ODE

```
...
void initialize(){
...

...

    H[0][0] = gsl_complex_add(H[0][0], MatPot);
```

We assume an electron number density of $N_e \approx 8.1 \times 10^{23}$ cm^{-3}, corresponding to the average number density of rock in the Earth's crust, as derived earlier. This value corresponds to an average mass density of $\rho_E \approx 2.7$ g/cm^3. As previously noted, matter effects become significant over propagation distances of the order of a thousand kilometres or more.

To illustrate the size of this effect, we solve the ODE and compute the electron neutrino appearance probability for the DUNE experiment's baseline of $L \approx 1300$ km, scanning over neutrino energies. We compare the oscillation probability $P(\nu_\mu \rightarrow \nu_e)$ for two cases: with only the vacuum term and with both the vacuum and matter terms included. In this example, we fix the CP-violating phase to $\delta_{CP} = -\pi/2$. The resulting curves are shown in Fig. 6.6. There is a clear enhancement in the appearance probability when matter effects are included, compared to the vacuum-only case. For antineutrinos, the matter potential changes sign, leading to an opposite effect. The size of this modification also depends on the neutrino energy, the true (currently unknown) value of δ_{CP}, and, importantly, on the choice of the mass hierarchy. These dependencies enhance the sensitivity of long-baseline experiments to the mass hierarchy, among other parameters. One way to visualize this sensitivity is through a so-called *biprobability* plot. For a fixed neutrino energy and a range of δ_{CP} values, one plots $P(\nu_\mu \rightarrow \nu_e)$ versus $P(\bar{\nu}_\mu \rightarrow \bar{\nu}_e)$, doing this separately for the normal and inverted mass hierarchies. The result, shown in Fig. 6.7, reveals that in the absence of matter effects (i.e. in vacuum), there would be a significant *degeneracy* between the two hierarchies, making it difficult to distinguish between them experimentally. However, the inclusion of matter effects breaks this degeneracy: The two scenarios separate cleanly in biprobability space. This enhances the capability of accelerator neutrino experiments to discriminate between the different mass hierarchy hypotheses, provided sufficient statistical precision is achieved.

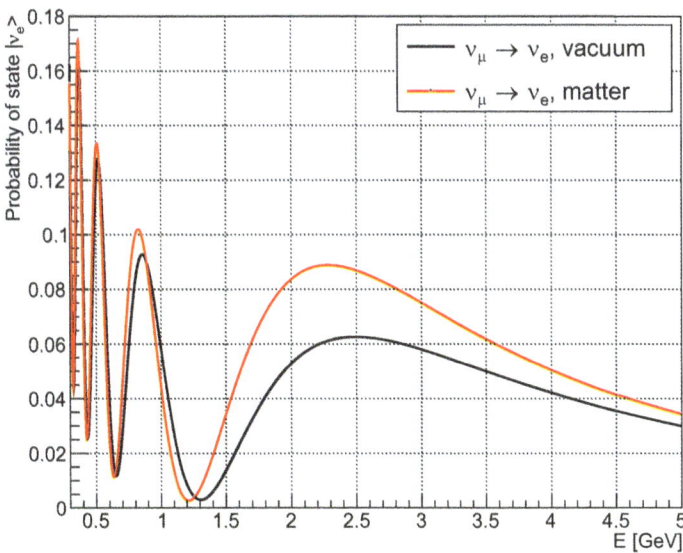

Fig. 6.6 Energy dependence of the three-flavour neutrino oscillation probability with a fixed propagation length of $L = 1300$ km, for electron neutrino appearance: $P(\nu_\mu \rightarrow \nu_e)$, comparing the vacuum-only case (black) and the case with matter effects included (red), assuming a rock density of $\rho_E = 2.7$ g/cm^3

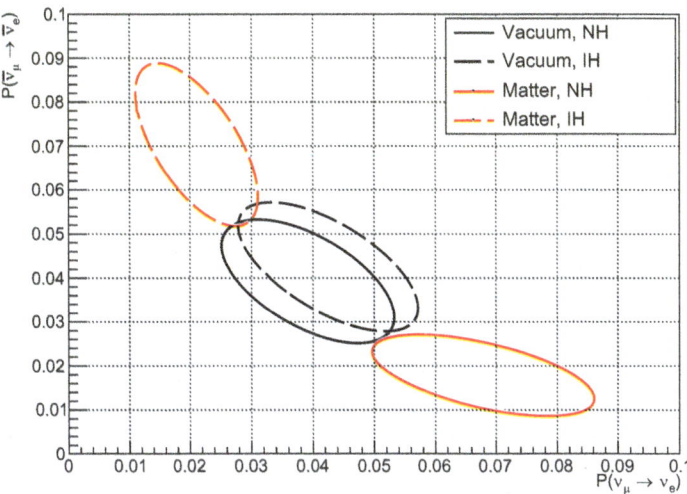

Fig. 6.7 Probability of electron neutrino appearance versus electron antineutrino appearance while scanning the value of δ_{CP} in the range $[-\pi, \pi]$, and assuming neutrino energy $E = 2$ GeV, baseline $L = 1300$ km, and rock density $\rho_E = 2.7$ g/cm^3. The biprobability is shown for normal (NO, solid) and inverted mass hierarchy (IO, dashed), vacuum only (black) and with the matter term included (red)

Another important manifestation of matter effects in a medium with constant density is the phenomenon of *resonant enhancement of oscillations* [15]. As discussed earlier, the presence of matter modifies the Hamiltonian of the system via an additional potential. This alters the effective mixing angles and mass-squared differences compared to their vacuum values. We define a quantity that is useful to express the relative dominance of vacuum vs. matter oscillations. In vacuum, the characteristic *oscillation length* is given by

$$l_\nu = \frac{4\pi E}{\Delta m^2},$$

which corresponds to the distance over which a full oscillation cycle—i.e. a phase change of 2π—occurs. In matter, a key quantity is the *refraction length*,

$$l_0 = \frac{\sqrt{2}\pi}{G_F N_e},$$

which is the distance required for the additional phase induced by matter effects to reach 2π. The term *refraction* refers to the fact that neutrino flavours experience different effective potentials due to charged-current interactions with electrons, with a potential difference of $V = V_1 - V_2 = \sqrt{2}G_F N_e$.

The effective mixing angles and mass-squared differences in matter depend, in general, on the ratio of the oscillation and refraction lengths:

$$x \equiv \frac{l_\nu}{l_0} \propto E N_e. \tag{6.25}$$

Mikheev and Smirnov discovered [16, 17] that for a certain value of x, a resonance occurs, resulting in a maximal transition probability between neutrino flavours in the presence of matter.

Reusing our existing code, it is straightforward to demonstrate this effect. We again assume the constant matter density of the Earth's crust and show that the resonance occurs at a particular value of x. The resonance condition can be written as

$$l_\nu = l_0 \cos 2\theta \approx 0.4\, l_0,$$

using the value of θ_{12}. In the code, this effect can be illustrated by computing the *survival probability* of electron neutrinos—that is, the probability that a neutrino initially in the state $|\nu_e\rangle$ remains in the same state after propagating a fixed distance:

$$P(\nu_e \to \nu_e)(L) = |\langle \nu_e| |\nu(L)\rangle|^2.$$

We scan the neutrino energy over a broad range to vary x and enhance the visibility of the effect by choosing a long propagation distance, $L = 10 l_0$. For example, at a

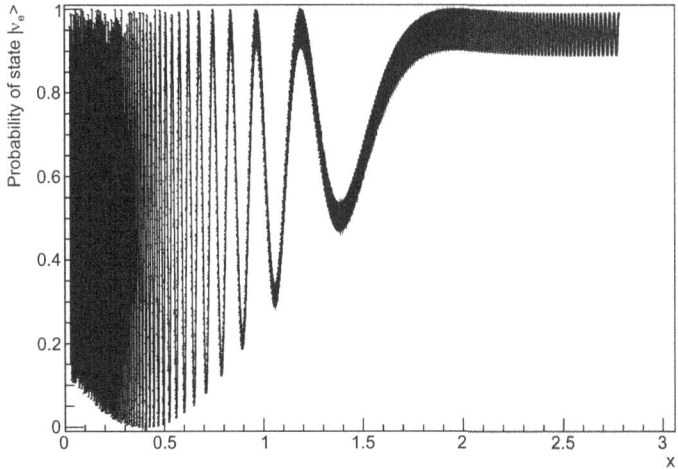

Fig. 6.8 Survival probability of electron neutrinos as a function of $x = l_\nu/l_0$ in a thick layer $L \approx 10 l_0$, showing a resonance enhancement of flavour oscillations in matter with constant density. The resonance maximum occurs at $x \approx 0.4$

neutrino energy of $E = 10$ MeV and assuming the Earth's crust electron density, we obtain $l_0 \approx 12{,}000$ km.

The result is shown in Fig. 6.8. On top of rapid, small-amplitude oscillations, the envelope of the oscillation probability exhibits a resonant peak at $x \approx 0.4$, with a clear attenuation at both lower and higher values of x. This signifies maximal mixing between electron neutrinos and other flavours due to their coherent forward scattering on electrons in the medium. In this example, since the matter density is constant, the effective mixing angles and mass-squared differences remain fixed during propagation, but the relative phases of the neutrino eigenstates evolve— driving the oscillation behaviour.

6.3.4 Propagation in Matter with Variable Density—the Sun

Having introduced the constant matter term in the neutrino propagation ODE, we can now extend the framework to include a spatially varying matter potential. A historically significant example where this variation plays a crucial role is the propagation of electron neutrinos produced in the core of the Sun.

In contrast to the previous case with constant density, a varying matter density leads to a time- (or distance-)dependent Hamiltonian, $H = H(t)$. As a consequence, the eigenstates of the Hamiltonian and the effective mixing angles evolve during propagation, in addition to the usual evolution of the relative phases between neutrino eigenstates. Depending on the local electron density and the neutrino energy, resonance conditions for flavour conversion may or may not be satisfied along the path.

As a concrete example, we can calculate the survival probability $P(\nu_e \rightarrow \nu_e)$ for electron neutrinos produced at the centre of the Sun and propagating radially outward, as a function of neutrino energy. To account for the varying density, we model the electron number density in the Sun using an exponential function [18]:

$$N_e(r) = N_e(0)\, e^{-r/r_0}, \qquad (6.26)$$

where $N_e(0) = 245\, N_A\ \mathrm{cm}^{-3}$ is the central electron density, and $r_0 = R_\odot/10.54$, with $R_\odot = 696{,}100\,\mathrm{km}$ the solar radius.

Recall that in Sect. 4.3 we calculated the energy spectrum of neutrinos from the proton-proton chain and we found that their range is $E \leq 20$ MeV. When one solves the ODE for few MeV neutrinos, the oscillations have very high frequency. Therefore, rather than using directly the survival probability, it is better to take the *average* survival probability, $\bar{P}(\nu_e \rightarrow \nu_e)$. Given the above parametrization, it is straightforward to solve the ODE from a pure ν_e initial state and obtain the following dramatic result: In the vacuum case, for solar neutrinos, independent of the neutrino energy, the average probability is $\bar{P}(\nu_e \rightarrow \nu_e)^{\mathrm{Sol}}_{\mathrm{vac}} \approx 0.6$. However, redoing the calculation with the addition of the varying solar electron density profile (matter effects) changes the picture: At $E \approx 20$ MeV, the survival probability $\bar{P}(\nu_e \rightarrow \nu_e)^{\mathrm{Sol}}_{\mathrm{mat}} \approx 0.3$, which is almost half the value of vacuum-only solution! The numerical solution from the ODE is illustrated in Figs. 6.9 and 6.10. Due to the high frequency of oscillation, we obtain bands of oscillations. The wide dark grey band shows the vacuum-only case. It is the same band in the background of both figures. The average value for the vacuum-only case can be eyeballed to be at the middle of the band, around $\bar{P} = 0.6$ for both plots. When we look at the light grey bands (solution with matter effects), we see that at $E = 20$ MeV the average value for the survival probability drops to around $\bar{P} = 0.3$ when exiting the Sun at $L \approx 700 \cdot 10^3$ km, while for $E = 5$ MeV the band starts to get wider, and as a result the average tends to increase. For even lower energy neutrinos, $E < 1$ MeV, the band would completely open and match the vacuum solution, and the average survival probability would be close to the vacuum value. The reader is encouraged to redo the same calculations. A possible code is shown below in code listing 6.6.

As a historical context, in 2002 the SNO collaboration announced that they solved the long-standing solar neutrino problem that the observed solar neutrino flux is much lower than that expected from theory.[10] The experiment demonstrated [19] that neutrinos from the Sun change their initial electron neutrino flavour state by the time they are detected at Earth. SNO measured neutral-current interactions of neutrinos on a D_2O target, and, since flavour oscillation effects do not have an impact in neutral-current interaction measurements, the corresponding observed solar flux matched the predicted value. Since then, Super-Kamiokande [20], Borexino [21], and other experiments have performed further measurements of the ^{8}B and ^{7}Be solar

---

[10] For a very nice overview of the solar neutrino problem, a suggested reading is J. N. Bahcall and R. Davis Jr., "Solar neutrinos: a scientific puzzle," Science, vol. 191, no. 4224, pp. 264–267, 1976.

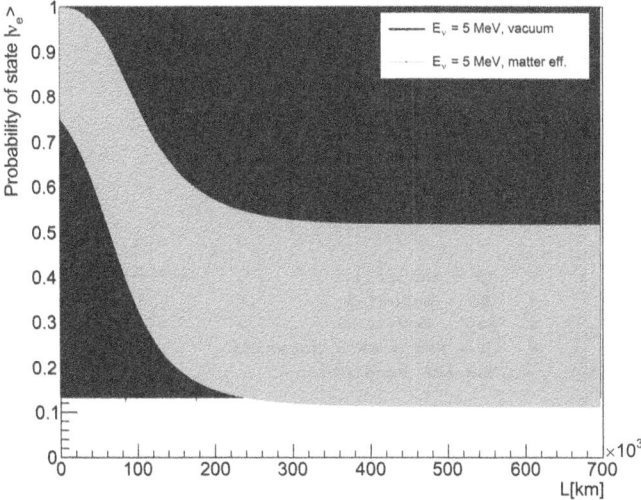

Fig. 6.9   Survival probability for $E = 5$ MeV electron neutrinos produced at the centre of the Sun as a function of the distance from the centre

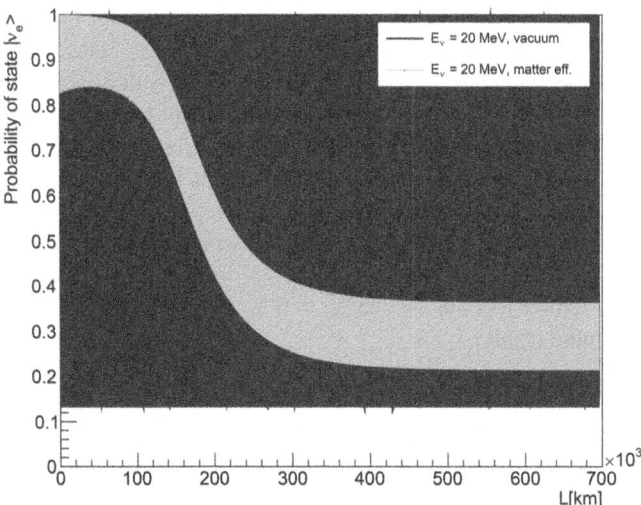

Fig. 6.10   Survival probability for $E = 20$ MeV electron neutrinos produced at the centre of the Sun as a function of the distance from the centre

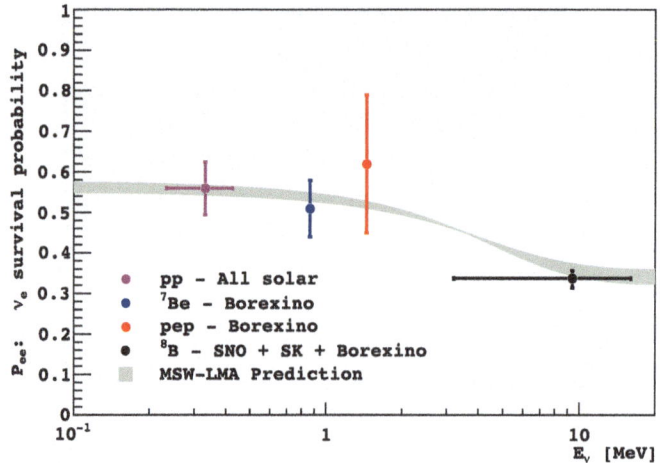

Fig. 6.11 Solar neutrino flux measurements performed by experiments [25]

neutrino fluxes. The SAGE [22], GALLEX [23], and GNO [24] groups identified the deficit in the solar neutrino flux from pp fusion. In addition, the neutrino oscillation parameters, Δm_{21}^2 and θ_{12}, have been fitted to the observed data. The solar neutrino deficit was observed to be energy dependent, and with the so-called MSW model[11] the effect could be explained by oscillation parameters with a large mixing angle solution. This allows confirmation of the prediction (that we just found out with our own code) that at low neutrino energies vacuum-dominated oscillation occurs in the Sun, while as the energy increases matter effects become significant and lead to lower survival probabilities and thus to a deficit when detected at Earth. See experimental results in Fig. 6.11, taken from [25], which confirms what we obtained solving the ODE: Above neutrino energies $E_\nu > 10$ MeV, the experimental value indeed becomes closer to $\bar{P}(\nu_e \rightarrow \nu_e)_{\text{mat}}^{\text{Sol}} \approx 0.3$, while at low energies the value is close to the vacuum result.

Code listing 6.6 ODE implementation for varying matter density profile from the Sun for a fixed energy

```
...
// 3x3 complex U, Uconj, and mass matrices
gsl_complex U[3][3], UconjT[3][3], Mass[3][3], H1[3][3], H[3][3];

// Instantaneous matter potential
gsl_complex MatPot_cur = gsl_complex_rect(0, 0);
```

[11] Named after Russian physicists S. P. Mikheyev (1940–2011), A. Smirnov (1951-), and American physicist L. Wolfenstein (1923–2015).

```
void initialize(){

  // exp(i*delta) and exp(-i*delta)
  gsl_complex idelta = gsl_complex_rect(0,delta);
  gsl_complex imdelta = gsl_complex_rect(0,-delta);
  gsl_complex exp_idelta = gsl_complex_exp(idelta);
  gsl_complex exp_imdelta = gsl_complex_exp(imdelta);

  // e1, e2, e3
  U[0][0] = gsl_complex_rect(c12*c13,0);
  U[0][1] = gsl_complex_rect(s12*c13,0);
  U[0][2] = gsl_complex_mul_real(exp_imdelta, s13);

  // mu1, mu2, mu3
  U[1][0] =
↪   gsl_complex_sub(gsl_complex_rect(-s12*c23,0),gsl_complex_mul_real(exp_
    idelta, c12*s23*s13));
  U[1][1] =
↪   gsl_complex_sub(gsl_complex_rect(c12*c23,0),gsl_complex_mul_real(exp_
    idelta, s12*s23*s13));
  U[1][2] = gsl_complex_rect(s23*c13,0);

  // tau1, tau2, tau3
  U[2][0] =
↪   gsl_complex_sub(gsl_complex_rect(s12*s23,0),gsl_complex_mul_real(exp_
    idelta, c12*c23*s13));
  U[2][1] = gsl_complex_sub(gsl_complex_rect(-c12*s23,
↪   0),gsl_complex_mul_real(exp_idelta, s12*c23*s13));
  U[2][2] = gsl_complex_rect(c23*c13,0);

  // Constants to get the correct dimensions (Mega-eV, femto-meter,
↪   kilometer)
  double p = 1.0/(2.0*hbarc*1.0e+06*1.0e-18*Enu*1.0e+09);
  // printf("p: %.5e\n", p);

  // calculate U conjugate transpose, and initialize H and M
  for(int i = 0; i < 3; i++){
    for(int j = 0; j < 3; j++){
      UconjT[i][j] = U[j][i];
      H[i][j] = gsl_complex_rect(0,0);
      H1[i][j] = gsl_complex_rect(0,0);
      Mass[i][j] = gsl_complex_rect(0,0);
    }
  }
}
```

```
// Transpose UconjT
for(int i = 0; i < 3; i++)
  for(int j = 0; j < 3; j++)
    UconjT[i][j] = gsl_complex_conjugate(UconjT[i][j]);

// Mass squared differences
Mass[0][0] = gsl_complex_rect(0,0);
Mass[1][1] = gsl_complex_rect(dm21, 0);
Mass[2][2] = gsl_complex_rect(dm21+dm32,0);

// Calculate H1= Mass*UconjT
for(int i = 0; i < 3; i++){
  for(int j = 0; j < 3; j++){
    gsl_complex s = gsl_complex_rect(0, 0);
    for (int k = 0; k < 3; k++){
      s = gsl_complex_add(s,gsl_complex_mul(Mass[i][k], UconjT[k][j]));
      //         printf("%.5e + i %.5e\n", GSL_REAL(s), GSL_IMAG(s));
    }
    //       printf("\n");
    H1[i][j] = s;
  }
}

// Calculate Hamiltonian H = U*H1
for(int i = 0; i < 3; i++){
  for(int j = 0; j < 3; j++){
    for (int k = 0; k < 3; k++)
      H[i][j] = gsl_complex_add(H[i][j],
↪  gsl_complex_mul(U[i][k],H1[k][j]));
    H[i][j] = gsl_complex_mul_real(H[i][j], p);
  }
}
}


// Define the complex ODE system
int complex_ode_func(double t, const double y[], double dydt[], void
↪  *params) {
  (void)(t); /* avoid unused parameter warning */
  gsl_complex psi_1 = gsl_complex_rect(y[0], y[1]);
  gsl_complex psi_2 = gsl_complex_rect(y[2], y[3]);
  gsl_complex psi_3 = gsl_complex_rect(y[4], y[5]);
```

```
gsl_complex imag = gsl_complex_rect(0, -1);

// Subtract the current value from H[0][0] to be updated
H[0][0] = gsl_complex_sub(H[0][0], MatPot_cur);

// Now add the Solar value
double Ne0 = 245*NA; // cm^{-3}
double r0 = Rsolar/10.54; // km
double r = t; // km
double Ne = Ne0*TMath::Exp(-r/r0);

// The Matter potential
gsl_complex MatPot =
gsl_complex_rect(sqrt(2)*GFermi*Ne/(hbarc*1.0e+06*1.0e-18),0);

// Save the current Matter Potential value
MatPot_cur = MatPot;

// Add the new value of the matter potential to H[0][0]
H[0][0] = gsl_complex_add(H[0][0], MatPot);

// The right-hand side of the diff. equation
// using the GSL algebraic functions
gsl_complex dpsi_1dt = gsl_complex_mul(H[0][0],psi_1);
dpsi_1dt = gsl_complex_add(dpsi_1dt, gsl_complex_mul(H[0][1],psi_2));
dpsi_1dt = gsl_complex_add(dpsi_1dt, gsl_complex_mul(H[0][2],psi_3));
gsl_complex dpsi_2dt = gsl_complex_mul(H[1][0],psi_1);
dpsi_2dt = gsl_complex_add(dpsi_2dt, gsl_complex_mul(H[1][1],psi_2));
dpsi_2dt = gsl_complex_add(dpsi_2dt, gsl_complex_mul(H[1][2],psi_3));
gsl_complex dpsi_3dt = gsl_complex_mul(H[2][0],psi_1);
dpsi_3dt = gsl_complex_add(dpsi_3dt, gsl_complex_mul(H[2][1],psi_2));
dpsi_3dt = gsl_complex_add(dpsi_3dt, gsl_complex_mul(H[2][2],psi_3));

// multiply by -i
dpsi_1dt = gsl_complex_mul(dpsi_1dt,imag);
dpsi_2dt = gsl_complex_mul(dpsi_2dt,imag);
dpsi_3dt = gsl_complex_mul(dpsi_3dt,imag);

dydt[0] = GSL_REAL(dpsi_1dt);
dydt[1] = GSL_IMAG(dpsi_1dt);
dydt[2] = GSL_REAL(dpsi_2dt);
dydt[3] = GSL_IMAG(dpsi_2dt);
dydt[4] = GSL_REAL(dpsi_3dt);
dydt[5] = GSL_IMAG(dpsi_3dt);
```

```
    return GSL_SUCCESS;
}

int main(void){

  // Step size in km
  double h = 1;
  // Absolute and relative errors
  const double epsAbs = 1.0e-06;
  const double epsRel = 1.0e-06;

  // Energy
  Enu = 0.005; // GeV

  // Baseline
  double L = 0.0; // km
  double L1 = Rsolar; // km

  // Save the intermediate step values
  std::vector<double> L_v, prob_psi1_v, prob_psi2_v, prob_psi3_v;

  // Define the GSL ODE system
  gsl_odeiv2_system sys = {complex_ode_func, NULL, 6, NULL};
  gsl_odeiv2_driver *d = NULL;

  // Call the initialize function
  initialize();

  // Initial value of the system:
  // RE(psi_1), IM(psi_1), RE(psi_2), IM(psi_2, RE(psi_3), IM(psi_3)
  double y[6] = {1.0, 0.0, 0.0, 0.0, 0.0, 0.0};

  // The driver needs the system, stepping algo,
  // the step size and the abs, rel tolerances
  d= gsl_odeiv2_driver_alloc_y_new(&sys, gsl_odeiv2_step_rk8pd, h,
↪  epsAbs, epsRel);
  // Status of GSL stepper
  int status = GSL_SUCCESS;

  // Stepping through the evolution of the ODE
  for (double Li = 0; Li <= L1; Li += h) {

    // Apply the ODE driver step
```

```
status = gsl_odeiv2_driver_apply(d, &L, Li, y);
//    status = gsl_odeiv2_driver_apply_fixed_step(d, &L, h, 1, y);

// Check for errors
if (status != GSL_SUCCESS) {
  printf("error, return value=%d\n", status);
  break;
}

if (status == GSL_SUCCESS){

  // Retrieve the current wave function values
  gsl_complex psi_1 = gsl_complex_rect(y[0],y[1]);
  gsl_complex psi_2 = gsl_complex_rect(y[2],y[3]);
  gsl_complex psi_3 = gsl_complex_rect(y[4],y[5]);

  // Save the probabilities at the intermediate step values
  prob_psi1_v.push_back(gsl_complex_abs2(psi_1));
  prob_psi2_v.push_back(gsl_complex_abs2(psi_2));
  prob_psi3_v.push_back(gsl_complex_abs2(psi_3));

  L_v.push_back(L);
}
} // End stepping over the ODE evolution

gsl_odeiv2_driver_free(d);

// Create a ROOT output file to draw the result in graphs
// onto a canvas object
TFile * ofile = new TFile("out_neut_mat_Solar.root", "RECREATE");
// The canvas
TCanvas *c1 = new TCanvas("c1", "c1",10,65,700,500);
c1->cd();
c1->SetGridx();
c1->SetGridy();
c1->SetFrameBorderMode(0);

TGraph * gr_psi1_t = new TGraph(L_v.size(), &L_v[0], &prob_psi1_v[0]);
gr_psi1_t->SetName("Pe");gr_psi1_t->SetLineColor(kRed);
gr_psi1_t->Draw("AL");
gr_psi1_t->GetXaxis()->SetTitle("L[km]");
gr_psi1_t->GetYaxis()->SetTitle("Probability of state |#nu_{e}>");
gr_psi1_t->GetYaxis()->SetRangeUser(-0.01, 1.01);
```

```
// Write the Canvas to the output ROOT file
c1->Write();
gr_psil_t->Write();
ofile->Close();

return 0;
}
```

6.3.5 Mass Eigenvalues in Matter

As mentioned earlier, during neutrino propagation through a medium with varying density, the instantaneous Hamiltonian evolves with position. Consequently, the effective mass eigenvalues of the system also change. Recall that the vacuum values we used were $\Delta m_{32}^2 \equiv m_2^2 = 2.437 \times 10^{-3}\,\text{eV}^2$ and $\Delta m_{21}^2 \equiv m_1^2 = 7.37 \times 10^{-5}\,\text{eV}^2$.

We now use GSL to compute the effective mass eigenvalues in matter and, after applying the transformation with the U mixing matrix,[12] obtain the corresponding flavour eigenvectors as a function of the electron number density N_e, scanned over the range $0 \le N_e \le 100\,N_A$, where N_A is Avogadro's constant.

The resulting effective mass eigenvalues $m_{M_1}^2$ and $m_{M_2}^2$ (the third eigenvalue $m_{M_3}^2$ is omitted for clarity) are shown in the left panel of Fig. 6.12 for neutrinos with

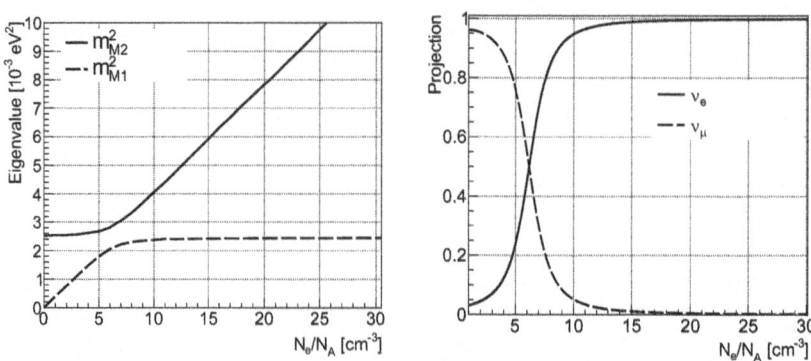

Fig. 6.12 Left: Effective mass eigenvalues, $m_{M_1}^2$ and $m_{M_2}^2$, of neutrinos with energy $E_\nu = 10$ MeV in matter as a function of the electron density. Right: Projection of the flavour eigenvectors onto the electron flavour basis state ν_e as a function of the electron density

---

[12] Note that, in general, the eigenvalues of a matrix A are invariant under a unitary transformation $U A U^{-1}$, but the eigenvectors are transformed by U.

energy $E_\nu = 10$ MeV. The right panel shows the projections of the corresponding flavour eigenvectors in matter onto the pure ν_e flavour basis state, $|\langle \Psi_e | \Psi_{x,M} \rangle|^2$.

These plots can be interpreted as follows. At large electron densities (e.g. the solar core), the effective mass eigenvalues are well separated, suppressing flavour transitions $\nu_e \leftrightarrow \nu_\mu$. The flavour eigenstates are approximately aligned with the mass eigenstates, with one eigenstate (ν_2) being nearly identical to the pure ν_e state. As the electron density decreases during outward propagation, the mass splitting between the eigenstates reaches a minimum at a specific density, corresponding to resonant flavour conversion. At this point, a near-complete $\nu_e \rightarrow \nu_\mu$ transition occurs. When the neutrino reaches regions of negligible density (e.g. outside the Sun), the mass eigenstate ν_2 becomes almost entirely composed of the ν_μ flavour, illustrating the MSW effect in action.

The code snippet in code listing 6.7 illustrates how to compute the eigenvalues and eigenvectors of a complex Hermitian matrix using GSL.

Code listing 6.7 Obtaining eigenvalues and eigenvectors of a complex Hermitian matrix in GSL

```
...
// Number of rows/columns
int m_Dim = 3;

// Allocate the GSL matrix object
gsl_matrix_complex* Hmat = gsl_matrix_complex_alloc (m_Dim, m_Dim);

// Reset matrix
gsl_matrix_complex_set_zero(Hmat);

// Set the GSL matrix elements to that of H
for (int n = 0; n < m_Dim;++n){
    for (int m = 0; m < m_Dim;++m){
        gsl_matrix_complex_set(Hmat, n, m, H[n][m]);
    }
}

// Allocate the objects for eigenvalues, vectors
gsl_vector *eval = gsl_vector_alloc (m_Dim);
gsl_matrix_complex *evec = gsl_matrix_complex_alloc (m_Dim, m_Dim);
gsl_eigen_hermv_workspace * w = gsl_eigen_hermv_alloc (m_Dim);

// Calculate the eigenvalues, eigenvectors
int out =  gsl_eigen_hermv (Hmat, eval, evec,  w);

// Obtain the first (real) eigenvalue
double eval_0 = gsl_vector_get (eval, 0);
```

```
// Obtain the corresponding (complex) eigenvector elements
gsl_complex evec_0 = gsl_matrix_complex_get(evec, 0, 0);
gsl_complex evec_1 = gsl_matrix_complex_get(evec, 1, 0);
gsl_complex evec_2 = gsl_matrix_complex_get(evec, 2, 0);
...
```

6.4   Neutrino Appearance Spectrum at a Long-Baseline Neutrino Experiment

In Chap. 5, we developed and implemented simplified event generators to model the phase-space kinematics of charged-current quasi-elastic (CCQE) and single-pion resonance (CC RES) scattering of neutrinos on nucleons, both with and without Fermi motion. Among these, CCQE is the dominant interaction channel at sub-GeV energies and plays a central role in some of the ongoing long-baseline neutrino oscillation experiments. This is because the energy of the incoming neutrino can be approximately reconstructed from the kinematics of the final-state particles.

In this section, we attempt to combine our simplified event generators with the neutrino flavour oscillation probability calculator—developed via numerical solution of the ODE—to approximately reproduce an important physical result observed in long-baseline neutrino experiments. The goal is not to produce a detailed simulation but rather to convey the essential idea of how such a result arises.

As a case study, we consider the Tokai-to-Kamioka (T2K) experiment, which was the first to report evidence for the appearance of electron neutrinos in a dominantly muon neutrino beam, $\nu_\mu \rightarrow \nu_e$ [26]. Our objective is to make a simplified prediction of the resulting electron neutrino energy spectrum in such a scenario. Because T2K was specifically designed to be sensitive to flavour oscillations over long baselines, we focus on the predicted energy spectrum at the far detector, Super-Kamiokande (SK). To sample the incoming neutrino energies somewhat realistically, we use the published SK flux prediction as an input to our simulation.

6.4.1   The T2K Experiment

The T2K experiment, sketched in Fig. 6.13, samples a narrowband, dominantly muon-neutrino beam produced at J-PARC, which is directed 2.5° off-axis towards the Super-Kamiokande (SK) water Cherenkov detector, located 295 km away. The off-axis configuration (see Sect. 4.4) results in a sharply peaked neutrino energy spectrum centred around $E \approx 0.6$ GeV. This setup is designed to suppress contributions from higher-energy neutrino interactions, such as deep inelastic scattering and higher-energy resonance production, enhancing sensitivity to CCQE processes.

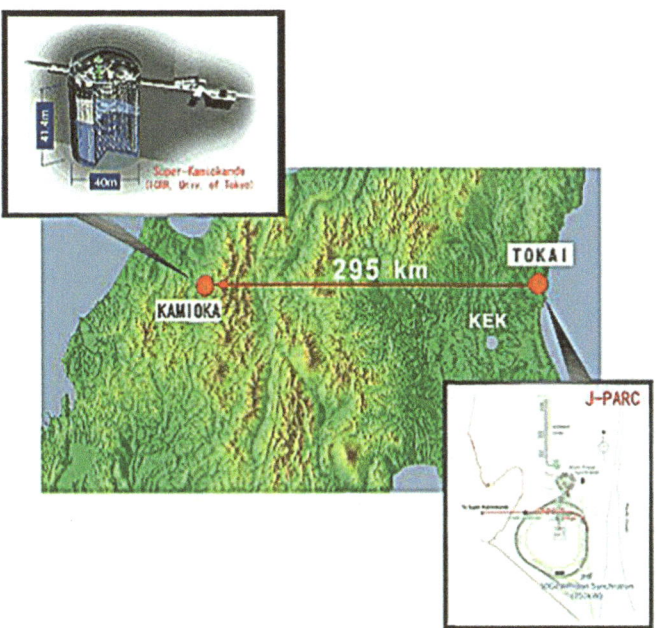

Fig. 6.13 Schematic overview of the T2K experiment, showing the neutrino beam trajectory (red arrow) and the locations of the near detector (ND280, at J-PARC near Tokai) and far detector (Super-Kamiokande, near Kamioka) over the map of Japan. Credit: the T2K experiment, http://t2k-experiment.org

The neutrino beam is sampled at two locations: first at the near detector (ND280), where measurements are made before significant flavour oscillations occur, and then at the far detector (Super-Kamiokande), where oscillation effects can be observed. Published flux predictions for both neutrino and antineutrino beams at the far detector are shown in Fig. 6.14.

In neutrino mode—also referred to as forward horn current (FHC) mode, based on the polarity of the magnetic horns used to focus the secondary pions and kaons—the T2K beam consists predominantly of muon neutrinos, with smaller fractions of muon antineutrinos, electron neutrinos, and electron antineutrinos. These subdominant components become relevant because the Super-Kamiokande detector is not magnetized, and therefore cannot distinguish the charge of the final-state leptons. As a result, antineutrino interactions can contribute to the observed signal, even in neutrino mode. The situation is reversed in antineutrino mode, or reverse horn current (RHC) mode, where the beam contains mainly antineutrinos but still includes a non-negligible neutrino contamination. Because of this, contributions from all possible neutrino and antineutrino species are typically simulated and taken into account when analysing the data. These components are treated as backgrounds when they originate from the "wrong-sign" part of the beam.

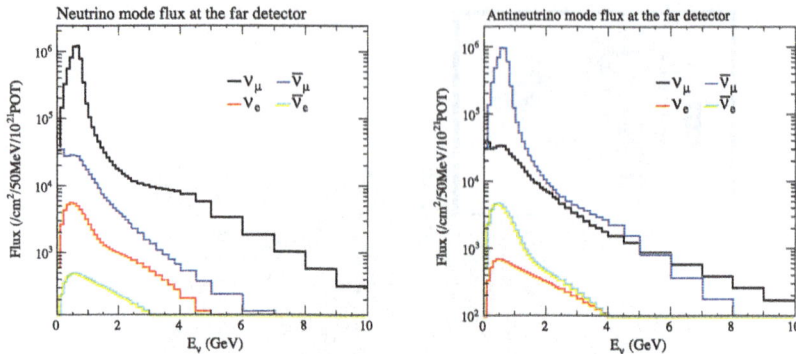

Fig. 6.14 Predicted unoscillated neutrino fluxes at the T2K Far Detector in neutrino (left) and antineutrino modes (right). The predictions are based on a tuning procedure to thin target hadron production data [27]

To constrain both the beam flux model and key neutrino interaction cross-section parameters (such as those governing CCQE and CC RES processes), the T2K experiment employs the ND280 near detector suite, shown in Fig. 6.15. Positioned close to the neutrino production point at J-PARC, ND280 observes the beam before significant flavour oscillation effects occur, and benefits from a much higher flux than at the far detector.

It is important to note that the neutrino flux is not a directly measurable quantity, but rather a model with associated uncertainties. Unlike charged leptons or protons, neutrinos interact so weakly with matter that their flux cannot be precisely characterized via direct measurement. Instead, the flux model is constructed through detailed simulations of the primary proton beam, target, magnetic horns, decay volume, and surrounding materials. These simulations rely heavily on external hadron production data and models and carry systematic uncertainties. The proximity of the ND280 detector allows for the collection of a large sample of neutrino-nucleus interactions. Using multiple nuclear targets and high-resolution tracking detectors, the experiment can reconstruct the final-state kinematics of each interaction with high precision. This, in turn, enables *constraints* on the beam flux, detector, and cross-section model uncertainties, improving the accuracy of predictions also at the far detector.

At the far site, neutrinos from the beam interact with the large water target of the Super-Kamiokande (SK) detector, a water Cherenkov detector with a fiducial volume of 22.5 kton, located in the Mozumi mine in Japan's Gifu prefecture [29]. SK has been operational since 1996 and continues to take data. It is instrumented with over 11,000 photomultiplier tubes (PMTs) that detect Cherenkov light emitted by charged particles produced in various interactions. An additional outer detector region, equipped with around 1,800 PMTs, serves as a veto to suppress cosmic ray background events. Super-Kamiokande has excellent particle identification capabilities and can reliably distinguish between electron-like and muon-like events based on the Cherenkov ring patterns and their opening angles.

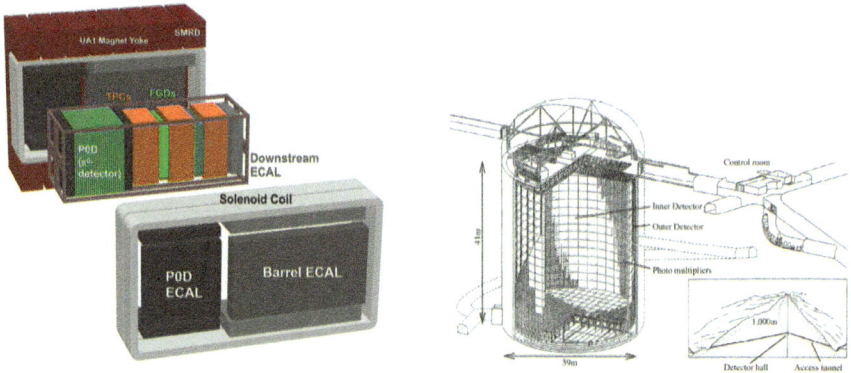

Fig. 6.15 Illustration of the T2K experiment's ND280 near detector complex (left)[28] and the Super-Kamiokande far detector (right). The source articles of the figures were published under Creative Common license: http://creativecommons.org/licenses/by/4.0/. Credit: the T2K experiment, http://t2k-experiment.org

A variety of interaction types can occur in SK when exposed to a neutrino beam. The most relevant are charged-current (CC) interactions, which produce a charged lepton in the final state. These charged leptons are used to identify the flavour of the incoming neutrino. When the energy of the outgoing lepton exceeds a threshold, it emits Cherenkov radiation, while traversing the water volume. The Cherenkov thresholds (in total energy) for electrons, muons, and pions are 0.768 MeV, 158.7 MeV, and 209.7 MeV, respectively. Cherenkov light is emitted in a cone along the path of the charged particle. This light is projected onto the walls of the detector, which are lined with PMTs, and appears as a ring pattern. Examples of rings produced by muons and electrons are shown in Fig. 6.16. The shape, intensity, and timing of the detected Cherenkov light allow for the reconstruction of the charged lepton's production vertex, momentum, and energy.

A comprehensive description of the T2K beam, near detectors, and far detector systems is provided in Ref. [27]. In what follows, we use only the central values of the published neutrino flux model, the target volume, material, and the reported event selection efficiencies as external inputs to the simulation of the appearance spectrum.

Charged-current quasi-elastic (CCQE) neutrino scattering events typically produce single-ring Cherenkov light events in Super-Kamiokande, since there is only one outgoing charged lepton—either an electron or a muon—in the final state.[13] Charged-current resonant (CC RES) events, on the other hand, can produce one or more additional mesons in the final state. These may fall below the Cherenkov

---

[13] Note that flavour oscillations can lead to the presence of tau neutrinos in the beam. However, the energy threshold for tau neutrinos to produce a tau lepton via charged-current scattering is around 3.5 GeV, which is well above the typical T2K neutrino energy. Tau neutrinos can also interact via neutral-current processes.

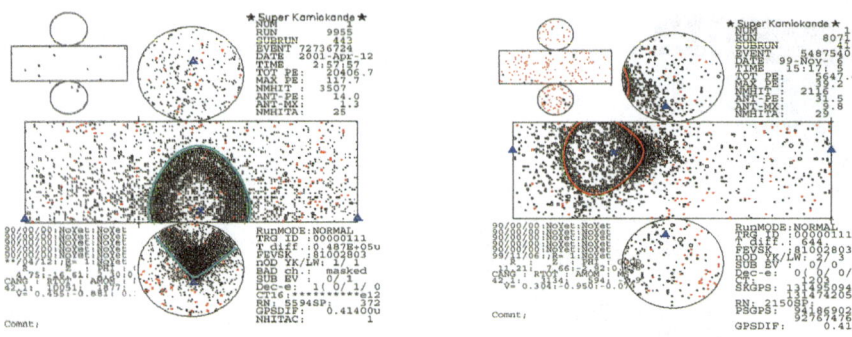

Fig. 6.16 Cherenkov rings from a muon (left) and electron (right) event in the Super-Kamiokande experiment [30]. The source article of the figures was published under Creative Common license: http://creativecommons.org/licenses/by/4.0/

threshold and go undetected, or their decay products may generate additional charged particles, leading to delayed or extra Cherenkov rings. Consequently, events in SK are classified based on the type of charged particle detected and the number of Cherenkov rings observed. Super-Kamiokande can reconstruct the momentum, energy, and direction of the outgoing charged lepton with high precision. This makes it possible to infer the energy of the incoming neutrino from the final-state kinematics using the reconstruction formula given in Eq. 5.43.

In the context of neutrino oscillation analysis, the goal is to isolate a sample enriched in CCQE events. To achieve this, selection criteria are applied to suppress events with multiple rings, which are more likely to arise from non-CCQE inter-actions. Nevertheless, some non-CCQE charged-current (CC) or neutral-current (NC) events inevitably pass the selection due to the high-energy tail in the neutrino flux. Because event classification is never perfect, there is always some level of contamination from misidentified events or misreconstructed particle properties.

After this initial discussion, we are now in a position to use the codes developed earlier—in Chap. 5—for CCQE and single-pion resonance event generation. These generators allow us to simulate neutrino-nucleon scattering events by sampling the initial neutrino energy from the T2K beam flux model, separately for each neutrino flavour component. The initial neutrino energy can be reconstructed from the final-state kinematics using the formulae derived previously. In addition, we can use the ODE solver introduced at the beginning of this chapter to compute the oscillation probabilities for various flavour transitions as a function of neutrino energy. These probabilities are then used to weight the generated scattering events accordingly.

It is important to clarify which beam flux components and oscillation proba-bilities are relevant for the different signal and background contributions. We use the muon (anti)neutrino beam flux both for the signal processes $\nu_\mu \rightarrow \nu_e$ and $\bar{\nu}_\mu \rightarrow \bar{\nu}_e$ (i.e. electron neutrino appearance), and for the background arising from muon (anti)neutrino survival, $\nu_\mu \rightarrow \nu_\mu$ and $\bar{\nu}_\mu \rightarrow \bar{\nu}_\mu$. For the beam components ν_e and $\bar{\nu}_e$, we apply the corresponding survival (disappearance) probabilities. The relevant cases are summarized in Table 6.2.

Table 6.2 Various contributions from the neutrino beam flux and oscillation channels to the electron neutrino appearance spectrum

Process	Flux component	Appearance | Disappearance
$\nu_\mu \to \nu_e$	ν_μ	Appearance (signal)
$\bar{\nu}_\mu \to \bar{\nu}_e$	$\bar{\nu}_\mu$	Appearance (signal)
$\nu_e \to \nu_e$	ν_e	Disappearance
$\bar{\nu}_e \to \bar{\nu}_e$	$\bar{\nu}_e$	Disappearance
$\nu_\mu \to \nu_\mu$	ν_μ	Disappearance
$\bar{\nu}_\mu \to \bar{\nu}_\mu$	$\bar{\nu}_\mu$	Disappearance

Once we have generated a sufficiently large number of events and assigned oscillation probability weights according to each of the processes listed in the table, we can construct the resulting neutrino energy spectra. These contributions can then be summed to form the full predicted spectrum. However, before doing so, we must address the issue of normalization—specifically, how to scale the spectra to match the accumulated statistics of a real dataset in a simplified way.

6.4.2 Event Generator Normalization

Suppose we have a CCQE or CC RES event generator at our disposal[14] and have used it to generate a sample of neutrino-nucleon scattering events. To make a meaningful prediction for the neutrino energy spectrum corresponding to each beam component, we must normalize the generated events to match the statistics of a given dataset.

Experimental data is typically quoted in terms of the number of protons-on-target (POT), and the neutrino beam flux $\Phi(E_\nu)$ (see Fig. 6.14) provides a reference for estimating the number of neutrinos passing through the detector volume per POT. The flux is given in units of [number of neutrinos / cm^2 / 50 MeV / 10^{21} POT] as a function of neutrino energy.

The total number of expected events in the detector is given by

$$N_{\text{exp}} = N_\nu \cdot \sigma_{\text{tot}} \cdot N_{\text{targets}}, \tag{6.27}$$

where:

- N_ν is the total number of neutrinos per cm^2 for a given amount of POT, obtained by integrating the beam flux.

---

[14] To simplify the discussion, we do not include neutral-current (NC) processes in this example. These processes can contribute to the appearance neutrino energy spectrum, primarily through π^0 production followed by decay into photons. However, their implementation—particularly for resonant production—follows the general approach of the Rein-Sehgal model, which we have already used for CC RES interactions.

- σ_{tot} is the total interaction cross section (in cm^2), estimated from the generated sample.
- $N_{targets}$ is the number of target nucleons in the detector.

To scale the generated events to this data, we assign a scaling factor to the events:

$$W = \frac{N_{exp}}{N_{gen}} = \frac{N_\nu \cdot \sigma_{tot} \cdot N_{targets}}{N_{gen}}, \tag{6.28}$$

where N_{gen} is the total number of generated events in a simulated sample.

To estimate N_ν, the total number of neutrinos per cm^2, we integrate the flux over energy and scale it to the desired POT. For instance, to scale to 1.5×10^{21} POT:

$$N_\nu = 1.5 \cdot 10^{21} \int \frac{\Phi(E_\nu)}{10^{21}} \, dE_\nu. \tag{6.29}$$

The total cross section of a sample can be estimated using the Monte Carlo integration formulas derived for rejection sampling in Sect. 2.2.2. In this context, the target function is the differential cross section evaluated for each event. For instance, in the Llewellyn Smith CCQE model, the differential cross section is expressed in terms of Q^2. Note that, since this is a two-to-two process, all phase-space weights are equal, and Q^2 is sampled uniformly by the event generator. In the case of the CC RES process, both W and Q^2 are also sampled uniformly. Therefore, we can directly apply the integral formulas derived earlier.

Next, we estimate the number of target nucleons in the detector. For SK, with a 22 kton water target, the number of nucleons is approximated by

$$N_{targets} \approx \rho \cdot V_{fid} \cdot N_A, \tag{6.30}$$

where $N_A = 6.02 \times 10^{23}$ mol^{-1} is Avogadro's constant. The number of atoms in 1 cm^3 of material is $N_{atoms} = \rho N_A / A$, where A is the atomic mass in g/mol. Since the number of nucleons per atom is approximately equal to the atomic mass A, the number of nucleons per unit volume is roughly $N_{nucleons} \approx \rho \cdot N_A$. For water with $\rho = 1$ g/cm^3, this gives

$$N_{nucleons} \approx 6.02 \cdot 10^{23} \text{ nucleons/cm}^3.$$

For a 22 kton fiducial mass, we obtain

$$N_{targets} \approx 1.35 \cdot 10^{34} \text{ nucleons}.$$

Finally, it is important to recognize that for a given beam mode, there are four flux components (muon, antimuon, electron, and antielectron neutrinos), several distinct interaction processes (CCQE, CC RES, etc.), and six possible oscillation channels (as shown in Table 6.2). Each case must be correctly normalized using the procedure above and then combined in a histogram to produce the final predicted neutrino energy spectrum.

6.4.3 The Electron Neutrino Appearance Spectrum

The result of the prediction is shown in Fig. 6.17 (left), obtained using only the codes developed in this book (with the exception of the far detector beam flux model, target volume, and selection efficiencies, as discussed). For comparison, the published result of the T2K experiment from 2021 is shown on the right [27]. To recap the steps we followed to obtain this spectrum:

- Neutrino energies were randomly sampled from the T2K far detector beam flux model for each neutrino flavour component.
- For each sampled initial-state neutrino energy, we generated CCQE and CC RES (anti)neutrino scattering events on nucleons using the Llewellyn Smith and Rein-Sehgal models, respectively, via rejection sampling and including Fermi motion to account for the initial-state nucleon momentum distribution.
- For each generated event, the reconstructed initial neutrino energy was computed from the final-state kinematics.
- Flavour oscillation probabilities—both appearance, $P(\nu_x \rightarrow \nu_y)$, and disappearance, $P(\nu_x \rightarrow \nu_x)$—were evaluated for every event and for each channel listed in Table 6.2, using the ODE solver for a baseline of 295 km.
- The reconstructed neutrino energy from each event was used to populate histograms, weighted by the corresponding oscillation probability, to produce the various oscillated spectra.
- Each spectrum was normalized to the target POT value of $1.5 \cdot 10^{21}$, following the procedure outlined in the previous section.

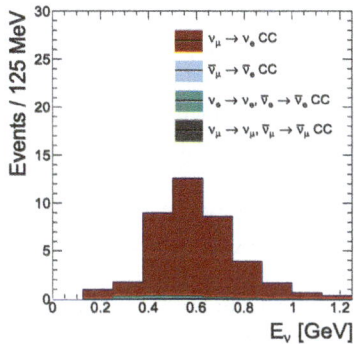 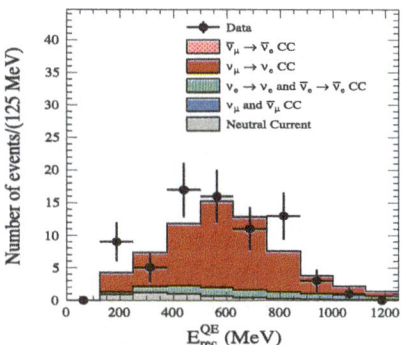

Fig. 6.17 Predicted appearance spectrum of electron neutrinos at the T2K Far Detector for $1.5 \cdot 10^{21}$ POT. Left: Result obtained using the event generators and flavour oscillation codes developed in this book. Right: Official result published by the T2K experiment [27]

- Finally, for each reweighted and normalized spectrum (as listed in Table 6.2), we applied a scaling factor based on the reported event selection efficiencies for single-ring electron events, as published by the T2K collaboration [27].

We observe very good qualitative agreement between our predicted neutrino spectrum and the official published result. The distribution is dominated by the $\nu_\mu \rightarrow \nu_e$ CC events in both cases, with background contributions mainly from $\nu_e \rightarrow \nu_e$ and $\bar{\nu}_e \rightarrow \bar{\nu}_e$ CC events. While the detailed shapes of the distributions differ— as expected—this is due to our intentional omission of detector and reconstruction effects (apart from the application of selection efficiencies) to keep the exercise manageable. Nevertheless, the normalization procedure, combined with efficiency scaling, yields a reasonable result. The readers are encouraged to try to redo the exercise on their own.

It is important to note that we have not simulated the background contributions from neutral-current (NC) events, which are known to be non-negligible. A further improvement would be to apply a true-to-reconstructed energy response matrix to our spectra, following the published figures in Super-Kamiokande [31]. Despite serious simplifications, we have successfully captured the essential features of one major result of the experiment and demonstrated how the expected appearance spectrum of a long-baseline neutrino experiment can be reproduced using the tools developed throughout this short guide book.

6.5   Future Experiments

Neutrino physics encompasses a wide range of past, ongoing, and future experiments that promise to deepen our understanding of neutrino properties, the structure of matter, and the fundamental symmetries of the universe. These efforts span a variety of experimental approaches, including long-baseline oscillation studies, reactor experiments, solar neutrino measurements, and astrophysical neutrino detection.

Future long-baseline experiments such as the Deep Underground Neutrino Experiment (DUNE) and Hyper-Kamiokande aim to probe CP violation in the lepton sector and to determine the neutrino mass hierarchy with unprecedented sensitivity. At the same time, reactor-based projects like JUNO will provide precise measurements of oscillation parameters and offer independent tests of the mass hierarchy. On the astrophysical frontier, experiments such as IceCube-Gen2 and KM3NeT are poised to extend our view into the high-energy universe, possibly revealing the sources of cosmic neutrinos. In parallel, projects like SNO+ and THEIA aim to unify solar, reactor, and double-beta decay physics through flexible detection capabilities. In addition, at the Large Hadron Collider $E_\nu \geq 200\,\text{GeV}$ neutrinos produced at 13.6 TeV centre-of-mass energy proton-proton collisions have recently been successfully detected by the FASER collaboration. As these and many more ambitious projects come online, they will bring both new challenges and exciting opportunities for research—making this a particularly thrilling time to be involved in the field of neutrino physics. Table 6.3 summarizes some of the

Table 6.3 Past, ongoing, and future neutrino experiments

Neutrino experiment	Type	Reference
Homestake	Solar	Astrophys. J. 496, 505 (1998)
GALLEX	Solar	Phys. Lett. B 447, 127 (1999)
SAGE	Solar	Phys. Atom. Nucl. 65, 2006 (2002)
IMB	Atmospheric, supernova	Phys. Rev. Lett. 58, 1494 (1987)
Kamiokande	Solar, atmospheric	Phys. Lett. B 280, 146 (1992)
Super-Kamiokande	Atmospheric, solar, oscillation	Phys. Rev. Lett. 81, 1562 (1998)
SNO	Solar	Phys. Rev. Lett. 89, 011301 (2002)
KamLAND	Reactor, geoneutrinos	Phys. Rev. Lett. 90, 021802 (2003)
K2K	Long-baseline oscillation	Phys. Rev. Lett. 90, 041801 (2003)
Chooz	Reactor	Eur. Phys. J. C 27, 331–374 (2003)
Double Chooz	Reactor	Phys. Rev. Lett. 108, 131801 (2012)
RENO	Reactor	Phys. Rev. Lett. 108, 191802 (2012)
Daya Bay	Reactor	Phys. Rev. Lett. 108, 171803 (2012)
MINOS	Long-baseline oscillation	Phys. Rev. Lett. 101, 131802 (2008)
T2K	Long-baseline oscillation	Nucl. Instrum. Meth. A 659, 106 (2011)
NOvA	Long-baseline oscillation	Phys. Rev. D 106, 032004 (2022)
BOREXINO	Solar, geoneutrinos	Phys. Rev. D 89, 112007 (2014)
LSND	Short-baseline, sterile search	Phys. Rev. D 64, 112007 (2001)
MINERvA	Neutrino-nucleus scattering	Phys. Rev. Lett. 111, 022501 (2013)
MicroBooNE	Short-baseline, sterile search	Phys. Rev. Lett. 128, 241801 (2022)
MiniBooNE	Short-baseline, sterile search	Phys. Rev. Lett. 121, 221801 (2018)
ICARUS	Short-baseline, sterile search	Nucl. Part. Phys. Proc. 300–302, 210 (2018)
DUNE	Long-baseline oscillation	DUNE Far Detector TDR II (arXiv:2002.03005)
JUNO	Reactor	J. Phys. G 43, 030401 (2016)
Hyper-Kamiokande	Long-baseline osc., solar, atm.	arXiv:2505.15019 (2025)
ANTARES	Atmospheric	Phys. Lett. B 816, 136228 (2021)
KM3NeT	Astrophysical, atmospheric	Nature 638, 376 (2025)
IceCube	Astrophysical	Phys. Rev. Lett. 111, 021103 (2013)
IceCube-Gen2	Astrophysical	J. Phys. G: Nucl. Part. Phys. 48, 060501 (2021)
SNO+	Double-beta dec., geo, reactor, solar	JINST 16 P08059 (2021)
ANNIE	Neutrino interaction studies	JINST 15 P03011 (2020).
THEIA	Multipurpose optical ν det.	Eur. Phys. J. C 80, 416 (2020)
ESSνSB	Spallation ν source	Eur. Phys. J. Spec. Top. 231, 3779–3955 (2022)
FASER	Collider neutrinos	Phys. Rev. D 110, 012009 (2024)
LiquidO	MeV neutrinos	Nature Com. Phys. 4, 273 (2021)
ENUBET	ν beam monitor	Eur. Phys. J. C 83, 964 (2023)
nuSTORM	ν from μs	arXiv:2505.06137 (2025)
NuTAG	ν tagging for long baseline	Eur. Phys. J. C 84, 1024 (2024)

major past, ongoing, and planned future experiments, their characteristics, and some references to publications.

The selected examples in this book were intended to illustrate the rich physics one can explore on their own computer with just a bit of effort. Computer simulations will continue to play a central role in the development and optimization of present and future experiments, from detector design and background modelling to data analysis and theoretical interpretation—and every discovery starts with someone poking at codes, plotting graphs, and telling themselves, "Let's just see what happens..."

References

1. R.L. Workman et al. (Particle Data Group), Prog. Theor. Exp. Phys. **2022**, 083C01 (2022) and 2023 update
2. S.F. King, Neutrino mixing: from experiment to theory. Nuclear Particle Phys. Proc. **265–266**, 288–295 (2015)
3. C. Jarlskog, Commutator of the quark mass matrices in the standard electroweak model and a measure of maximal CP nonconservation. Phys. Rev. Lett. **55**, 1039 (1987)
4. L. Wolfenstein, Neutrino oscillations in matter. Phys. Rev. D **17**, 2369 (1978)
5. S.P. Mikheyev, A.Y. Smirnov, 3ν oscillations in matter and solar neutrino data. Phys. Lett. B **200**, 560–564 (1988)
6. L.L. Chau, W.Y. Keung, Comments on the parametrization of the Kobayashi-Maskawa matrix. Phys. Rev. Lett. **53**, 1802 (1984)
7. Z. Maki, M. Nakagawa, S. Sakata, Remarks on the unified model of elementary particles. Progr. Theor. Phys. **28**(5), 870–880 (1962)
8. B. Pontecorvo, Neutrino experiments and the problem of conservation of leptonic charge. Sov. Phys. JETP **26**, 984–988 (1968)
9. I. Esteban, M.C. Gonzalez-Garcia, M. Maltoni, T. Schwetz, A. Zhou, The fate of hints: updated global analysis of three-flavour neutrino oscillations. JHEP **2020**, 178 (2020)
10. A.K. Ichikawa, Design concept of the magnetic horn system for the T2K neutrino beam. NIM A **690**, 27–33 (2012)
11. K. Abe et al. (T2K Collaboration), Evidence of electron neutrino appearance in a muon neutrino beam. Phys. Rev. D **88**, 032002 (2013)
12. K. Abe et al. (Hyper-Kamiokande Collaboration), Physics potential of a long-baseline neutrino oscillation experiment using a J-PARC neutrino beam and Hyper-Kamiokande. Prog. Theor. Exp. Phys. **2015**, 053C02 (2015)
13. R. Acciarri et al. (DUNE Collaboration), Long-baseline neutrino facility (LBNF) and deep underground neutrino experiment (DUNE) conceptual design report volume 2: The Physics program for DUNE at LBNF (2016). arXiv:1512.06148v2 [physics.ins-det]
14. K. Abe et al. (T2K Collaboration), Constraint on the matter–antimatter symmetry-violating phase in neutrino oscillations. Nature **580**, 339–344 (2020)
15. A.Y. Smirnov, The MSW effect and solar neutrinos (2003). arXiv:0305106 [hep-ph]
16. S.P. Mikheev, A.Y. Smirnov, Resonant amplification of neutrino oscillations in matter and spectroscopy of solar neutrinos. Sov. J. Nucl. Phys. **42**, 913–917 (1985)
17. S.P. Mikheev, A.Y. Smirnov, Resonant amplification of neutrino oscillations in matter and solar neutrino spectroscopy. Nuovo Cim. **C9**, 17–26 (1986)
18. C. Giunti, *Fundamentals of Neutrino Physics* (Oxford University Press, Oxford, 2007)
19. Q.R. Ahmad et al. (SNO Collaboration), Direct evidence for neutrino flavor transformation from neutral-current interactions in the Sudbury neutrino observatory. Phys. Rev. Lett. **89**, 011301 (2002)

20. K. Abe et al. (Super-Kamiokande Collaboration), Solar neutrino results in Super-Kamiokande-III. Phys. Rev. D **83**, 052010 (2011)
21. C. Arpesella et al. (Borexino Collaboration), Direct measurement of the ^7Be Solar neutrino flux with 192 days of Borexino data. Phys. Rev. Lett. **101**, 091302 (2008)
22. J.N. Abdurashitov et al. (SAGE Collaboration), Solar neutrino flux measurements by the Soviet-American gallium experiment (SAGE) for half the 22-year solar cycle. J. Exp. Theor. Phys. **95**, 181–193 (2002)
23. W. Hampel et al. (GALLEX Collaboration), GALLEX solar neutrino observations: results for GALLEX IV Phys. Lett. B **447**, 127–133 (1999)
24. M. Altmann et al. (GNO Collaboration), Complete results for five years of GNO solar neutrino observations. Phys. Lett. B **661**, 174–190 (2005)
25. V. Antonelli, L. Miramonti, C. Peña Garay, A. Serenelli, Solar neutrinos. Adv. High Energy Phys. **2013**, 351926 (2020)
26. K. Abe et al. (T2K Collaboration), Indication of electron neutrino appearance from an accelerator-produced off-axis muon neutrino beam. Phys. Rev. Lett **107**, 041801 (2011)
27. K. Abe et al. (T2K Collaboration). Improved constraints on neutrino mixing from the T2K experiment with 3.13×10^{21} protons on target. Phys. Rev. D **103**, 112008 (2021). https://doi.org/10.1103/PhysRevD.103.112008
28. K. Abe et al. (T2K Collaboration), Measurements of neutrino oscillation parameters from the T2K experiment using 3.6×10^{21} protons on target. Eur. Phys. J. C **83**, 782 (2023)
29. S. Fukuda et al. (Super-Kamiokande Collaboration), The Super-Kamiokande detector. Nucl. Instrum. Methods Phys. Res. Sect. A **501**, 418 (2003)
30. T. Nakaya, K. Nishikawa, Long baseline neutrino oscillation experiments with accelerators in Japan. Eur. Phys. J. C **80**, 344 (2020)
31. E. Richard et al. (Super-Kamiokande Collaboration), Measurements of the atmospheric neutrino flux by Super-Kamiokande: Energy spectra, geomagnetic effects, and solar modulation. Phys. Rev. D **94**, 052001 (2016)

Appendix

A

A.1 Derivation of the 2- and 3-Body Phase-Space Factors

The general form for the n-body Lorentz-invariant phase-space factor is given as follows [1,2]:

$$R_n = \int d^4 p_1' \int d^4 p_2' \cdots \int d^4 p_n'$$

$$\times \delta^4(p_1' + p_2' + \cdots + p_n' - p_1 - p_2)$$

$$\times \prod_{i=1}^{n} \delta(p_i'^2 - m_i^2). \tag{A.1}$$

Starting with the above definition, the two-body phase-space factor can be written following the definition in Eq. A.1. Let us express it in the rest frame of a decaying particle, $P \to p_1 + p_2$, with $P = (M, \mathbf{0})$ and mass M:

$$R_2(M; p_1, p_2) = \int d^4 p_1 \int d^4 p_2 \delta^4(p_1 + p_2 - P)\delta(p_1^2 - m_1^2)\delta(p_2^2 - m_2^2). \tag{A.2}$$

Using the Dirac-delta integral, we can evaluate the p integrals using $d^4 p \delta(p^2 - m^2) \to d^3 p / 2E$:

$$R_2(M; p_1, p_2) = \int \frac{d^3 p_1 d^3 p_2}{2E_1 E_2} \delta^4(p_1 + p_2 - P)$$

$$= \int \frac{d^3 p_1 d^3 p_2}{2E_1 E_2} \delta^3(\mathbf{p}_1 + \mathbf{p}_2)\delta(M - \sqrt{\mathbf{p}_1^2 + m_1^2} - \sqrt{\mathbf{p}_2^2 + m_2^2}). \tag{A.3}$$

© The Author(s), under exclusive license to Springer Nature Switzerland AG 2026   195
B. Radics, *Neutrino Physics*, Lecture Notes in Physics 1043,
https://doi.org/10.1007/978-3-032-03993-4

Since $\mathbf{p}_1 = -\mathbf{p}_2$, we can simplify the $\delta^3(\mathbf{p}_1 + \mathbf{p}_2)$ integral. Let us integrate over p_2 and obtain

$$R_2(M; p_1, p_2) = \int \frac{4\pi p^2 dp}{4\sqrt{p^2 + m_1^2}\sqrt{p^2 + m_2^2}} \delta(M - \sqrt{p^2 + m_1^2} - \sqrt{p^2 + m_2^2}),$$

(A.4)

where we have defined $|\mathbf{p}_1| = |\mathbf{p}_2| = p$, rewritten d^3p as $4\pi p^2 dp$, having performed angular integrations since in the rest frame of P the decay is isotropic, and used $E = \sqrt{p^2 + m^2}$.

In order to finish the derivation, we need to rewrite the remaining $\delta(g(p))$ function and express p. Let us use the four-momentum conservation, $P = p_1 + p_2$. The four-momenta of the final-state particles is $p_i = (\mathbf{p}_i, E_i)$, $i = 1, 2$. Because we are in the rest frame of the decaying particle, we find

$$Pp_1 = ME_1, \quad Pp_2 = ME_2 \tag{A.5}$$

$$E_1 = \frac{Pp_1}{M}, \quad E_2 = \frac{Pp_2}{M}.$$

But we also have the following relationship, again using the four-momentum conservation

$$p_1 p_2 = \frac{1}{2}[(p_1 + p_2)^2 - p_1^2 - p_2^2] = \frac{1}{2}(M^2 - m_1^2 - m_2^2). \tag{A.6}$$

Then we can express the energy and the magnitude of the three-momenta of the final-state particles using the invariant masses. Note again that in the rest frame of the decaying particle the two final-state particles must go out back to back, $\mathbf{p}_1 = -\mathbf{p}_2$:

$$E_1 = \frac{Pp_1}{M} = \frac{(p_1 + p_2)p_1}{M} = \frac{M^2 + (m_1^2 - m_2^2)}{2M} \tag{A.7}$$

$$E_2 = \frac{Pp_2}{M} = \frac{(p_1 + p_2)p_2}{M} = \frac{M^2 - (m_1^2 - m_2^2)}{2M}$$

$$|\mathbf{p}|^2 = |\mathbf{p}_1|^2 = |\mathbf{p}_2|^2 = E_2^2 - m_2^2 = \frac{M^4 - 2M^2(m_1^2 - m_2^2) + (m_1^2 - m_2^2)^2 - 4M^2 m_2^2}{4M^2}$$

$$= \frac{M^4 - 2M^2(m_1^2 + m_2^2) + (m_1^2 - m_2^2)^2}{4M^2}$$

$$= \frac{[M^2 - (m_1 + m_2)^2][M^2 - (m_1 - m_2)^2]}{4M^2}.$$

In the last line, we obtained the magnitude of the three-momenta of the daughter particles, $|\mathbf{p}| \equiv p$, in the rest frame of the parent particle. To further evaluate Eq. A.4, we use the following identity of the Dirac-delta function for function composition

$$\delta(g(x)) = \frac{\delta(x - x_0)}{|g'(x)|_{x=x_0}},$$ (A.8)

where it is assumed that $g(x)$ has a root in x_0. Identifying $x = p'$ and using Eq. A.7, we have

$$\left. \frac{dg}{dp'} \right|_{p'=p} = p \left(\frac{1}{\sqrt{p^2 + m_1^2}} + \frac{1}{\sqrt{p^2 + m_2^2}} \right) = \frac{Mp}{\sqrt{p^2 + m_1^2}\sqrt{p^2 + m_2^2}}.$$ (A.9)

The last equation can be derived from energy conservation. In the rest frame of the decaying particle, we have $M = E_1 + E_2 = \sqrt{p^2 + m_1^2} + \sqrt{p^2 + m_2^2}$. From this, we can write

$$M = E_1 + E_2$$

$$\frac{1}{E_1} + \frac{1}{E_2} = \frac{M}{E_1 E_2}$$

$$p \left(\frac{1}{E_1} + \frac{1}{E_2} \right) = \frac{Mp}{E_1 E_2}.$$

Now we go back to Eq. A.4 and carry on with the integral and obtain the 2-body phase-space factor in the rest frame of a decaying particle

$$R_2(M; p_1, p_2) = \frac{\pi \sqrt{p^2 + m_1^2}\sqrt{p^2 + m_2^2}}{Mp} \int \frac{p'^2 dp'}{\sqrt{p'^2 + m_1^2}\sqrt{p'^2 + m_2^2}} \delta(p' - p)$$

$$= \frac{\pi p(M; m_1, m_2)}{M}$$

$$= \frac{\pi}{2M^2} \sqrt{[M^2 - (m_1 + m_2)^2][M^2 - (m_1 - m_2)^2]}.$$ (A.10)

We can express the 3-body phase-space factor for the decay $P \rightarrow p_1 + p_2 + p_3$ as a product of two factors: (1) a phase-space factor describing the decay $P \rightarrow P_{1,2} + p_3$ with $P_{1,2}$ representing a composite "particle" $P_{1,2} = p_1 + p_2$, and (2) a phase-space factor describing the "decay" of the composite "particle" $P_{1,2} \rightarrow p_1 + p_2$, with $M_{1,2} = \sqrt{P_{1,2}^2} = \sqrt{(p_1 + p_2)^2}$

$$R_3(M; m_1, m_2, m_3) = \frac{\pi^2}{M M_{1,2}} p(M; M_{1,2}, m_3) \times p(M_{1,2}; m_1, m_2).$$ (A.11)

References

1. James, F, *Monte Carlo Phase Space*. https://cds.cern.ch/record/275743 (CERN 1968)
2. Hagedorn, R, *Relativistic Kinematics* (W. A. Benjamin, 1964)

Index

© The Editor(s) (if applicable) and The Author(s), under exclusive license to
Springer Nature Switzerland AG 2026
B. Radics, *Neutrino Physics*, Lecture Notes in Physics 1043,
https://doi.org/10.1007/978-3-032-03993-4

The manufacturer's authorised representative in the EU is Springer
Nature Customer Service Centre GmbH, Europaplatz 3, 69115 Heidelberg,
Germany. If you have any concerns regarding our products, please
contact ProductSafety@springernature.com

Printed and bound by CPI Group (UK) Ltd, Croydon, CR0 4YY

23/04/2026

02095598-0006